Carlos S. Kubrusly

Hilbert Space Operators

A Problem Solving Approach

Birkhäuser
Boston • Basel • Berlin

Carlos S. Kubrusly
Catholic University of Rio de Janeiro
R. Marques de S. Vicente 255
22453-900, Rio de Janeiro, RJ
Brazil

Library of Congress Cataloging-in-Publication Data

Kubrusly, Carlos S., 1947-
    Hilbert space operators : a problem solving approach / Carlos S. Kubrusly.
        p. cm.
    Includes bibliographical references and index.
    ISBN 0-8176-3242-5 (alk. paper) – ISBN 3-7643-3242-5 (alk. paper)
        1. Operator theory. 2. Hilbert space. I. Title

    QA329.K82 2003
    515'.724–dc21   2003052078
                                                                        CIP

AMS Subject Classifications: 47-XX, 47-01, 47-02, 47Axx, 47A05, 47A10, 47A12, 47A15, 47A16, 47A20, 47A62, 47A63, 47A75, 47Bxx, 47B10, 47B15, 47B20, 47B37, 47B47, 47B48

Printed on acid-free paper.
©2003 Birkhäuser Boston

Birkhäuser

ISBN 0-8176-3242-5   SPIN 10929520
ISBN 3-7643-3242-5

Typeset by the author.

9 8 7 6 5 4 3 2 1

Birkhäuser Boston • Basel • Berlin
A member of BertelsmannSpringer Science+Business Media GmbH

To Jessica and Alan

Preface

This is a problem book on Hilbert space operators (i.e., on *bounded linear* transformations of a Hilbert space into itself) where theory and problems are investigated together.

We treat only a part of the so-called single operator theory. Selected problems, ranging from standard textbook material to points on the boundary of the subject, are organized into twelve chapters. The book begins with elementary aspects of *Invariant Subspaces* for operators on Banach spaces in Chapter 1. Basic properties of *Hilbert Space Operators* are introduced in Chapter 2, *Convergence and Stability* are considered in Chapter 3, and *Reducing Subspaces* is the theme of Chapter 4. Primary results about *Shifts* on Hilbert space comprise Chapter 5. These are introductory chapters where the majority of the problems consist of auxiliary results that prepare the ground for the next chapters. Chapter 6 deals with *Decompositions* for Hilbert space contractions, Chapter 7 focuses on *Hyponormal Operators*, and Chapter 8 is concerned with *Spectral Properties* of operators on Banach and Hilbert spaces. The next three chapters (as well as Chapter 6) carry their subjects from an introductory level to a more advanced one, including some recent results. Chapter 9 is about *Paranormal Operators*, Chapter 10 covers *Proper Contractions*, and Chapter 11 searches through *Quasi-reducible Operators*. The final Chapter 12 commemorates three decades of *The Lomonosov Theorem* on nontrivial hyperinvariant subspaces for compact operators. The table of contents assigns a keyword to each problem. This keyword function is not injective over the whole book but its restriction to each chapter is.

The prerequisite for this book is a first course in elementary functional analysis. The reader is supposed to be familiar with basic definitions and fundamental results on normed and inner product spaces—in particular, on Banach and Hilbert spaces. For instance, I will often refer to [32] for these preliminary notions but any similar text would be equally suitable. From this point on the book is largely self-contained.

Each chapter surveys the background material, and problems are naturally posed as the theory develops. The collection of problems consists of propositions (whose solutions in fact are proofs), examples or counterexamples. Detailed solutions to all stated problems close each chapter. Once in a while the reader will find a question. These are open questions indicating specific topics of current research. In spite of this, the book is not intended to be a research monograph. Basic problems and auxiliary results, needed to solve further problems, are also proposed throughout the text. The difficulty of these problems increases gradually within each chapter. Many of them are the accomplishment of a large number of mathematicians over a long period. I have not attempted to trace back the origin of those problems; perhaps such an attempt would hardly be successful. However, some problems dealing with recent results are properly referred.

Among the diverse audience for whom the book is intended, the main target is the population of graduate students. By graduate students I mean students of mathematics, and also of engineering, economics, statistics and physics, who have already acquired the essential knowledge described above. The idea is to challenge active readers to build up part of the theory by themselves. Moreover, it can be useful for working mathematicians going into operator theory, and for scientists wishing to apply operator theory to their field.

In writing a problem book on Hilbert space operators, one is bound to be largely influenced by P. Halmos's great classic *A Hilbert Space Problem Book* [22]. Since I have been working on this subject for a long time, I also benefited from the help of many friends, among students and colleagues, and I am really grateful to all of them. Special thanks are due to my colleagues B.P. Duggal, L. Kérchy, N. Levan, A.C. Gadelha Vieira and P.C.M. Vieira. Their sensible comments motivated and clarified several parts of this book. I am grateful to Catholic University of Rio de Janeiro for providing the release time that made this project possible, to Birkhäuser Boston for a pleasant and lasting partnership, as well as to CNPq (Brazilian National Research Council) for a research grant.

Carlos S. Kubrusly
Rio de Janeiro
June 2003

Contents

9   Paranormal Operators   93

10   Proper Contractions   109

Hilbert Space Operators

1
Invariant Subspaces

Let $\mathcal{X}$ and $\mathcal{Y}$ be normed spaces and let $\mathcal{B}[\mathcal{X}, \mathcal{Y}]$ denote the normed space of all bounded linear transformations of $\mathcal{X}$ into $\mathcal{Y}$. Recall that "continuous linear transformation" and "bounded linear transformation" are synonyms (a transformation T of $\mathcal{X}$ into $\mathcal{Y}$ is *bounded* if there exists a constant $\beta \geq 0$ such that $\|Tx\| \leq \beta \|x\|$ for every x in $\mathcal{X}$). We shall use the same notation for the norms on $\mathcal{X}$, $\mathcal{Y}$ and also for the induced (uniform) norm on $\mathcal{B}[\mathcal{X}, \mathcal{Y}]$:

$$\|T\| = \sup_{\|x\| \leq 1} \|Tx\| = \sup_{\|x\| < 1} \|Tx\| = \sup_{\|x\| = 1} \|Tx\| = \sup_{x \neq 0} \frac{\|Tx\|}{\|x\|}$$

for every $T \in \mathcal{B}[\mathcal{X}, \mathcal{Y}]$ (the last two expressions hold for $\mathcal{X} \neq \{0\}$). By a *subspace* of a normed space $\mathcal{X}$ we mean a *closed* linear manifold of $\mathcal{X}$. The *kernel* (or *null space*) of $T \in \mathcal{B}[\mathcal{X}, \mathcal{Y}]$ is the inverse image of $\{0\}$ under T:

$$\mathcal{N}(T) = T^{-1}(\{0\}) = \{x \in \mathcal{X} : Tx = 0\},$$

which is a subspace of $\mathcal{X}$. The image of $\mathcal{X}$ under T,

$$\mathcal{R}(T) = T(\mathcal{X}) = \{y \in \mathcal{Y} : y = Tx \text{ for some } x \in \mathcal{X}\},$$

is the *range* of $T \in \mathcal{B}[\mathcal{X}, \mathcal{Y}]$, which is a linear manifold of $\mathcal{Y}$. If $\mathcal{X} = \mathcal{Y}$, then we put $\mathcal{B}[\mathcal{X}] = \mathcal{B}[\mathcal{X}, \mathcal{X}]$ for short. The elements of $\mathcal{B}[\mathcal{X}]$ are called *operators*. In other words, by an *operator* we mean a bounded linear transformation of a normed space $\mathcal{X}$ into itself, so that $\mathcal{B}[\mathcal{X}]$ is the normed algebra of all operators on $\mathcal{X}$. If $\mathcal{X} \neq \{0\}$, then $\mathcal{B}[\mathcal{X}]$ contains the identity operator I and $\|I\| = 1$, which means that $\mathcal{B}[\mathcal{X}]$ is a *unital* normed algebra. Recall that $\mathcal{B}[\mathcal{X}, \mathcal{Y}]$ is a Banach space whenever $\mathcal{Y}$ is a Banach space so that $\mathcal{B}[\mathcal{X}]$ is a unital Banach algebra whenever $\mathcal{X} \neq \{0\}$ is a Banach space.

Let $T \in \mathcal{B}[\mathcal{X}]$ be an operator on a normed space $\mathcal{X}$. A subset M of $\mathcal{X}$ is *T-invariant* (or M is an *invariant subset* for T) if $T(M) \subseteq M$ (i.e., if Tx lies in M whenever x lies in M). An *invariant linear manifold* (*invariant subspace*) for T is a linear manifold (subspace) of $\mathcal{X}$ that, as a subset of $\mathcal{X}$, is T-invariant. It is clear that the zero space $\{0\}$ and the whole space $\mathcal{X}$ are invariant subspaces for every T in $\mathcal{B}[\mathcal{X}]$.

Problem 1.1. Let $\mathcal{X}$ be a normed space, let M be a subset of $\mathcal{X}$, let M be a linear manifold of $\mathcal{X}$, and take an operator $T \in \mathcal{B}[\mathcal{X}]$. Show that

(a) if M is T-invariant, then its closure M^- also is T-invariant;

(b) if M is an invariant linear manifold for T, then M^- is an invariant subspace for T.

Lat$(\mathcal{X})$ will denote the collection of all subspaces of a normed space $\mathcal{X}$ (which, in fact, is a complete lattice in the inclusion ordering — see e.g., [32, p. 213]). The subspaces $\{0\}$ and $\mathcal{X}$ are the trivial elements of Lat$(\mathcal{X})$: a subspace in Lat$(\mathcal{X})$ is nontrivial if it is a proper nonzero subspace of $\mathcal{X}$ (i.e., $M \in$ Lat$(\mathcal{X})$ is nontrivial if $\{0\} \neq M \neq \mathcal{X}$). If dim $\mathcal{X} \leq 1$, then $\mathcal{X}$ has no nontrivial subspace: there exist nontrivial subspaces in Lat$(\mathcal{X})$ if and only if the dimension of $\mathcal{X}$ is greater than 1 (i.e., Lat$(\mathcal{X}) \neq \{\{0\}, \mathcal{X}\}$ if and only if dim $\mathcal{X} > 1$).

Consider the algebra $\mathcal{B}[\mathcal{X}]$ of all operators on a normed space $\mathcal{X}$ and take any operator T in $\mathcal{B}[\mathcal{X}]$. A *nontrivial invariant subspace* for T is a nontrivial element of Lat$(\mathcal{X})$ that is invariant for T (i.e., an $M \in$ Lat$(\mathcal{X})$ such that $\{0\} \neq M \neq \mathcal{X}$ and $T(M) \subseteq M$). An element of $\mathcal{B}[\mathcal{X}]$ is a *scalar operator* if it is a multiple of the identity, say, αI for some scalar α. It is clear that every subspace of $\mathcal{X}$ is invariant for any scalar operator. That is, every subspace in Lat$(\mathcal{X})$ is invariant for any scalar operator in $\mathcal{B}[\mathcal{X}]$, and so every scalar operator has a nontrivial invariant subspace if dim $\mathcal{X} > 1$.

A linear manifold (or a subspace) M of a normed space $\mathcal{X}$ is *hyperinvariant* for T in $\mathcal{B}[\mathcal{X}]$ if it is invariant for every operator in $\mathcal{B}[\mathcal{X}]$ that commutes with T. In other words, a linear manifold (or a subspace) M is hyperinvariant for $T \in \mathcal{B}[\mathcal{X}]$ if M is L-invariant for every $L \in \mathcal{B}[\mathcal{X}]$ such that $LT = TL$. Obviously, every hyperinvariant linear manifold (or subspace) for T is invariant for T.

Problem 1.2. Let $\mathcal{N}(T)$ and $\mathcal{R}(T)$ denote kernel and range, respectively, of an operator T on a normed space $\mathcal{X}$. Prove the following assertions.

(a) $\mathcal{N}(T)$ and $\mathcal{R}(T)^-$ are hyperinvariant subspaces for T.

(b) If dimension of $\mathcal{X}$ is greater than 1 and T has no nontrivial invariant subspace, then $\mathcal{N}(T) = \{0\}$ and $\mathcal{R}(T)^- = \mathcal{X}$.

Hint: $\mathcal{N}(T)$ and $\mathcal{R}(T)$ are linear manifolds of $\mathcal{X}$ because T is linear. Moreover, since T is bounded (i.e., continuous), $\mathcal{N}(T)$ is closed in $\mathcal{X}$. Therefore, $\mathcal{N}(T)$ and $\mathcal{R}(T)^-$ are subspaces (closed linear manifolds) of the normed space $\mathcal{X}$ (see Solution 1.1(b)). Also, if $F: \mathcal{X} \to \mathcal{X}$ is a continuous mapping, then $F(A^-) \subseteq F(A)^-$ for every subset A of $\mathcal{X}$ (see e.g., [32, p. 182]).

Recall that a linear transformation between linear spaces, say $L: \mathcal{X} \to \mathcal{Y}$, is injective if and only if $\mathcal{N}(L) = \{0\}$. A bounded linear transformation of a normed space into a normed space is *quasiinvertible* (or a *quasiaffinity*) if it is injective and has a dense range. Thus what Problem 1.2(b) says is: *if an operator has no nontrivial subspace, then it is quasiinvertible.*

Problem 1.3. Let S and T be nonzero operators on a normed space $\mathcal{X}$. If $ST = O$, then $\mathcal{N}(S)$ and $\mathcal{R}(T)^-$ are nontrivial invariant subspaces for both S and T.

Problems 1.2(a) and 1.3 ensure that, if the product of two nonzero operators is null, then both of them have nontrivial hyperinvariant subspaces, and they also have common nontrivial invariant subspaces.

Problem 1.4. Let $T \in \mathcal{B}[\mathcal{X}]$ be a nonzero operator acting on a normed space $\mathcal{X}$. If the operator equation $STS = TS$ has a nontrivial solution (i.e., a solution $S \in \mathcal{B}[\mathcal{X}]$ such that $O \neq S \neq I$), then T has a nontrivial invariant subspace. Prove.

Recall that an operator T is *nilpotent* if $T^n = O$ for some positive integer n, and *algebraic* if $p(T) = O$ for some nonzero polynomial p.

Problem 1.5. Let $\mathcal{X}$ be a normed space of dimension greater than 1 and show that

(a) a nonzero nilpotent operator has a nontrivial hyperinvariant subspace;

(b) if $\mathcal{X}$ is a *complex* normed space, then every nonscalar algebraic operator has a nontrivial hyperinvariant subspace.

Take an operator T in $\mathcal{B}[\mathcal{X}]$. The *commutant* of T is the set $\{T\}'$ consisting of all operators in $\mathcal{B}[\mathcal{X}]$ that commute with T:

$$\{T\}' = \{L \in \mathcal{B}[\mathcal{X}]: LT = TL\}.$$

Thus Problem 1.5(b) ensures that on a complex normed space *every operator in the commutant of a nonscalar algebraic operator has a nontrivial invariant subspace.* Equivalently, *if an operator has no nontrivial invariant subspace, then there is no nonscalar algebraic operator in its commutant.*

It is readily verified that $\{T\}'$ is an operator algebra that contains the identity. That is, $\{T\}'$ is a unital subalgebra of the normed algebra $\mathcal{B}[\mathcal{X}]$. Indeed, $\{T\}'$ is a linear manifold of the linear space $\mathcal{B}[\mathcal{X}]$, contains the identity of $\mathcal{B}[\mathcal{X}]$, and the product of any two elements of $\{T\}'$ lie in $\{T\}'$. Let $\mathcal{P}(T)$ denote the set of all polynomials of T. This is again an operator algebra that contains the identity; in fact, a unital subalgebra of $\{T\}'$.

Problem 1.6. Let T be an operator acting on a normed space. Show that the subspaces $\mathcal{N}(L)$ and $\mathcal{R}(L)^-$ are T-invariant for every $L \in \{T\}'$. In particular, the subspaces $\mathcal{N}(p(T))$ and $\mathcal{R}(p(T))^-$ are T-invariant for every polynomial p; that is, for every $p(T) \in \mathcal{P}(T)$.

Let $\mathrm{Lat}(T)$ denote the collection of *all invariant subspaces for* $T \in \mathcal{B}[\mathcal{X}]$, where $\mathcal{X}$ is a normed space. ($\mathrm{Lat}(T)$ is a complete lattice in the inclusion ordering.) Thus an operator T has no nontrivial invariant subspace if and only if $\mathrm{Lat}(T) = \{\{0\}, \mathcal{X}\}$. Take any operator T in $\mathcal{B}[\mathcal{X}]$ and an arbitrary vector x in $\mathcal{X}$, and consider the $\mathcal{X}$-valued power sequence $\{T^n x\}_{n=0}^\infty$. The range of this sequence (i.e., the set $\{T^n x\}_{n \geq 0}$) is called the *orbit of* x *under* T. Recall that the (linear) span of any subset M of a linear space, $\mathrm{span}\, M$, is the linear manifold made up of all (finite) linear combinations of vectors in M. Thus the (linear) span of the orbit of x under T is the set of the images of all nonzero polynomials of T at x; that is,

$$\mathrm{span}\{T^n x\}_{n \geq 0} = \{p(T)x \in \mathcal{X}\colon p \text{ is a nonzero polynomial}\}.$$

Since $\mathrm{span}\{T^n x\}_{n \geq 0}$ is a linear manifold of $\mathcal{X}$, it follows that its closure,

$$\bigvee\{T^n x\}_{n \geq 0} = \left(\mathrm{span}\{T^n x\}_{n \geq 0}\right)^-,$$

is a subspace of $\mathcal{X}$. Moreover, it is clear that $\mathrm{span}\{T^n x\}_{n \geq 0}$ is T-invariant, and nonzero whenever $x \neq 0$, and so is its closure. That is, $\bigvee\{T^n x\}_{n \geq 0}$ is a nonzero invariant subspace for T whenever $x \neq 0$. These are the *cyclic subspaces*: $\mathcal{M}$ in $\mathrm{Lat}(T)$ is cyclic for T if $\mathcal{M} = \bigvee\{T^n x\}_{n \geq 0}$ for some x in $\mathcal{X}$. If $\bigvee\{T^n x\}_{n \geq 0} = \mathcal{X}$ for some x in $\mathcal{X}$, then x is said to be a *cyclic vector* for T. If T has a cyclic vector, then it is a *cyclic operator*. We say that a linear manifold $\mathcal{M}$ of $\mathcal{X}$ is *totally cyclic* for T if every nonzero vector in $\mathcal{M}$ is cyclic for T. Observe that T has no nontrivial invariant subspace if and only if every nonzero vector in $\mathcal{X}$ is a cyclic vector for T (reason: if $\mathcal{M}$ is T-invariant, then $T^n(\mathcal{M}) \subseteq \mathcal{M}$); that is, if and only if $\bigvee\{T^n x\}_{n \geq 0} = \mathcal{X}$ for every $x \neq 0$ in $\mathcal{X}$; which means that $\mathcal{X}$ is itself totally cyclic for T.

Let $\mathcal{X}$ and $\mathcal{Y}$ be normed spaces. A bounded linear transformation X in $\mathcal{B}[\mathcal{X}, \mathcal{Y}]$ *intertwines* an operator $T \in \mathcal{B}[\mathcal{X}]$ to an operator $S \in \mathcal{B}[\mathcal{Y}]$ if

$$XT = SX.$$

In this case, a trivial induction shows that

$$XT^n = S^n X$$

for every nonnegative integer n, and hence

$$Xp(T) = p(S)X$$

for every polynomial p. If there exists a nonzero X intertwining T to S, then we say T is *intertwined* to S. If there exists an X with dense range intertwining T to S, then we say that T is *densely intertwined* to S.

Problem 1.7. Suppose $T \in \mathcal{B}[\mathcal{X}]$ is densely intertwined to $S \in \mathcal{B}[\mathcal{Y}]$ and let $X \in \mathcal{B}[\mathcal{X}, \mathcal{Y}]$ be a transformation with dense range intertwining T to S. If $x \in \mathcal{X}$ is a cyclic vector for T, then $Xx \in \mathcal{Y}$ is a cyclic vector for S. Consequently, if a linear manifold $\mathcal{M}$ of $\mathcal{X}$ is totally cyclic for T, then the linear manifold $X(\mathcal{M})$ of $\mathcal{Y}$ is totally cyclic for S.

Here is a sufficient condition for transferring nontrivial invariant subspaces from S to T whenever T is densely intertwined to S.

Problem 1.8. If T in $\mathcal{B}[\mathcal{X}]$ is densely intertwined to S in $\mathcal{B}[\mathcal{Y}]$, say, if

$$XT = SX \quad \text{with} \quad \mathcal{R}(X)^- = \mathcal{Y}$$

for some X in $\mathcal{B}[\mathcal{X}, \mathcal{Y}]$, then the inverse image $X^{-1}(\mathcal{M})$ of a nontrivial invariant subspace $\mathcal{M}$ for S is a nontrivial invariant subspace for T, provided that

$$\mathcal{R}(X) \cap \mathcal{M} \neq \{0\}.$$

It is worth noticing that the hypothesis $\mathcal{R}(X) \cap \mathcal{M} \neq \{0\}$ of Problem 1.8 is not redundant. Indeed, let $\mathcal{R}$ be a linear manifold of an arbitrary normed space $\mathcal{Y}$ of dimension greater than 1. If $\dim \mathcal{R} < \infty$, then $\mathcal{R}^- = \mathcal{Y}$ implies $\mathcal{R} = \mathcal{Y}$ (finite-dimensional linear manifolds are closed) and so $\mathcal{R} \cap \mathcal{M} \neq \{0\}$ whenever $\mathcal{M} \neq \{0\}$. But if $\dim \mathcal{R} = \infty$ and $\mathcal{R}^- = \mathcal{Y}$, then, of course, it may happen that $\mathcal{R} \neq \mathcal{Y}$ and, in this case, there exists a nonzero y in $\mathcal{Y} \backslash \mathcal{R}$. Put $\mathcal{M} = \text{span}\{y\}$, a nontrivial subspace of $\mathcal{Y}$ ($\mathcal{M}$ is a one-dimensional linear manifold of $\mathcal{Y}$). Since $\mathcal{R} \cap \mathcal{M}$ is a linear manifold of $\mathcal{Y}$ (intersection of linear manifolds is a linear manifold), it follows that if $\mathcal{R} \cap \mathcal{M} \neq \{0\}$, then $\mathcal{M} \subseteq \mathcal{R}$ (because $\dim \mathcal{M} = 1$) so that y lies in $\mathcal{R}$, which is a contradiction. Therefore, $\mathcal{R} \cap \mathcal{M} = \{0\}$.

However, if the intertwining X is surjective (i.e., if $\mathcal{R}(X) = \mathcal{Y}$), then the condition $\mathcal{R}(X) \cap \mathcal{M} \neq \{0\}$ is trivially satisfied whenever $\mathcal{M} \neq \{0\}$, and hence $X^{-1}(\mathcal{M})$ is a nontrivial invariant subspace for T whenever $\mathcal{M}$ is a nontrivial invariant subspace for S.

Two operators are *similar* if they are intertwined by an invertible bounded linear transformation with a bounded inverse. That is, T in $\mathcal{B}[\mathcal{X}]$ and S in $\mathcal{B}[\mathcal{Y}]$ are similar if $XT = SX$ (equivalently, $X^{-1}S = TX^{-1}$) for some invertible X in $\mathcal{B}[\mathcal{X}, \mathcal{Y}]$ such that X^{-1} lies in $\mathcal{B}[\mathcal{Y}, \mathcal{X}]$. Therefore, since X intertwines T to S and X^{-1} intertwines S to T, and since X and X^{-1} are bounded, linear and surjective, it follows as a particular case of Problem 1.8, that *if two operators are similar, and if one of them has a nontrivial invariant subspace, then so has the other.* In other words, similarity preserves nontrivial invariant subspaces.

Recall that the commutant $\{T\}'$ of an operator T in $\mathcal{B}[\mathcal{X}]$ is the unital algebra consisting of all operators that commute with T, which turns out to be the unital subalgebra of $\mathcal{B}[\mathcal{X}]$ of all operators intertwining T to itself. Also recall that a linear manifold (or a subspace) of a normed space $\mathcal{X}$ is hyperinvariant for T in $\mathcal{B}[\mathcal{X}]$ if it is invariant for every L in $\{T\}'$. Now take any operator T on a normed space $\mathcal{X}$ and, for each vector x in $\mathcal{X}$, set

$$\mathcal{T}_x = \{Lx \in \mathcal{X} : L \in \{T\}'\} = \bigcup_{L \in \{T\}'} Lx \subseteq \mathcal{X}.$$

Problem 1.9. Show that each $\mathcal{T}_x^-$ is a hyperinvariant subspace for T.

Obviously, $\mathcal{T}_0 = \{0\}$. Since $I \in \{T\}'$, it follows that x lies in $\mathcal{T}_x$. Hence $\mathcal{T}_x^- \neq \{0\}$ for every nonzero x in $\mathcal{X}$. Therefore, if T has no nontrivial hyperinvariant subspace, then $\mathcal{T}_x^- = \mathcal{X}$ for every $x \neq 0$.

Problem 1.10. Let $\mathcal{X}$ and $\mathcal{Y}$ be normed spaces. Take T in $\mathcal{B}[\mathcal{X}]$, S in $\mathcal{B}[\mathcal{Y}]$, X in $\mathcal{B}[\mathcal{X}, \mathcal{Y}]$ and Y in $\mathcal{B}[\mathcal{Y}, \mathcal{X}]$ such that

$$XT = SX \quad \text{and} \quad YS = TY.$$

Suppose $\mathcal{M}$ is a nontrivial hyperinvariant subspace for S. If

$$\mathcal{R}(X)^- = \mathcal{Y} \quad \text{and} \quad \mathcal{N}(Y) \cap \mathcal{M} = \{0\},$$

then $Y(\mathcal{M}) \neq \{0\}$ and, for each nonzero x in $Y(\mathcal{M})$, $\mathcal{T}_x^-$ is a nontrivial hyperinvariant subspace for T.

If the linear transformation Y is injective (i.e., if $\mathcal{N}(Y) = \{0\}$), then the condition $\mathcal{N}(Y) \cap \mathcal{M} = \{0\}$ is trivially satisfied. Thus, if $XT = SX$ with $\mathcal{R}(X)^- = \mathcal{Y}$ and $YS = TY$ with $\mathcal{N}(Y) = \{0\}$ (i.e., if T is densely intertwined to S and S is intertwined to T by an injective transformation), then T has a nontrivial hyperinvariant subspace whenever S has.

Recall that a bounded linear transformation X of a normed space $\mathcal{X}$ into a normed space $\mathcal{Y}$ is quasiinvertible (or a quasiaffinity) if it is injective and has a dense range (i.e., if $\mathcal{N}(X) = \{0\}$ and $\mathcal{R}(X)^- = \mathcal{Y}$). An operator T

in $\mathcal{B}[\mathcal{X}]$ is a *quasiaffine transform* of an operator S in $\mathcal{B}[\mathcal{Y}]$ if there exists a quasiinvertible transformation X in $\mathcal{B}[\mathcal{X}, \mathcal{Y}]$ intertwining T to S. Two operators are *quasisimilar* if they are quasiaffine transforms of each other. In other words, T in $\mathcal{B}[\mathcal{X}]$ and S in $\mathcal{B}[\mathcal{Y}]$ are quasisimilar if there exist X in $\mathcal{B}[\mathcal{X}, \mathcal{Y}]$ and Y in $\mathcal{B}[\mathcal{Y}, \mathcal{X}]$ such that

$$\mathcal{N}(X) = \{0\}, \quad \mathcal{R}(X)^- = \mathcal{Y}, \quad \mathcal{N}(Y) = \{0\}, \quad \mathcal{R}(Y)^- = \mathcal{X},$$

$$XT = SX \quad \text{and} \quad YS = TY.$$

Therefore, as a particular case of Problem 1.10, *if two operators are quasisimilar and if one of them has a nontrivial hyperinvariant subspace, then so has the other.* That is, quasisimilarity preserves nontrivial hyperinvariant subspaces. It is clear that similar operators are quasisimilar, but similarity does more than this; similarity also preserves nontrivial invariant subspaces.

Question: Does quasisimilarity preserve nontrivial invariant subspaces?

Solutions

Solution 1.1. Let T be an operator on $\mathcal{X}$ and let M be a subset of $\mathcal{X}$.

(a) Take an arbitrary x in M^- so that x is a point of adherence of M, and hence there exists an M-valued sequence, say $\{x_n\}$, that converges to x. If M is T-invariant, then $\{Tx_n\}$ is again an M-valued sequence. Since T is continuous, $\{Tx_n\}$ converges to Tx. But M^- is closed and each Tx_n lies in $M \subseteq M^-$ so that Tx lies in M^-. Indeed, the Closed Set Theorem (see e.g., [32, p. 118]) says that *a set A in a metric space X is closed if and only if every A-valued sequence that converges in X has its limit in A*). Therefore, M^- is T-invariant whenever M is.

(b) Let M be a linear manifold of the normed space $\mathcal{X}$. It follows by (a) that M^- is T-invariant whenever M is. But the closure of a linear manifold is again a linear manifold. Indeed, if x and y are arbitrary vectors in M^-, then there exist M-valued sequences $\{x_n\}$ and $\{y_n\}$ that converge to x and y, respectively, in the norm topology of $\mathcal{X}$. Since addition is continuous in the norm topology, it follows that the M-valued sequence $\{x_n + y_n\}$ converges to $x + y$, and hence $x + y$ lies in M^- (by the Closed Set Theorem). Similarly, αx lies in M^- for every scalar α whenever x lies in M^-. Thus M^- is a linear manifold, and hence a closed linear manifold; that is, a subspace.

Solution 1.2. Let T and L be operators on a normed space $\mathcal{X}$.

(a) Suppose L commutes with T. If x lies in $\mathcal{N}(T)$, then $TLx = LTx = 0$, and hence Lx lies in $\mathcal{N}(T)$. Thus $L(\mathcal{N}(T)) \subseteq \mathcal{N}(T)$ so that $\mathcal{N}(T)$ is

L-invariant. Since $LTx = TLx$ for every vector x in $\mathcal{X}$, it follows that $L(\mathcal{R}(T)) \subseteq \mathcal{R}(T)$, and hence $L(\mathcal{R}(T)^-) \subseteq L(\mathcal{R}(T))^- \subseteq \mathcal{R}(T)^-$ so that $\mathcal{R}(T)^-$ is L-invariant.

(b) If $\dim \mathcal{X} > 1$ and T has no nontrivial invariant subspace, then (obviously) T is nonzero and has no nontrivial hyperinvariant subspace, so that $\mathcal{N}(T)$ and $\mathcal{R}(T)^-$ are trivial subspaces by assertion (a). But, since $T \neq O$, it follows that $\mathcal{N}(T) \neq \mathcal{X}$ and $\mathcal{R}(T) \neq \{0\}$. Therefore, $\mathcal{N}(T) = \{0\}$ and $\mathcal{R}(T)^- = \mathcal{X}$.

Solution 1.3. If $ST = O$, then $\mathcal{R}(T) \subseteq \mathcal{N}(S)$, and hence

$$T(\mathcal{N}(S)) \subseteq T(\mathcal{X}) = \mathcal{R}(T) \subseteq \mathcal{N}(S).$$

If $T \neq O$, then $\mathcal{R}(T) \neq \{0\}$ so that $\mathcal{N}(S) \neq \{0\}$. If $S \neq O$, then $\mathcal{N}(S) \neq \mathcal{X}$ and so $\mathcal{R}(T)^- \neq \mathcal{X}$ because $\mathcal{N}(S)$ is closed. Therefore,

$$\{0\} \neq \mathcal{N}(S) \neq \mathcal{X} \quad \text{and} \quad \{0\} \neq \mathcal{R}(T)^- \neq \mathcal{X}.$$

Moreover, since S is continuous and since $S(\mathcal{R}(T)) = \{0\}$,

$$S(\mathcal{R}(T)^-) \subseteq S(\mathcal{R}(T))^- \subseteq \mathcal{R}(T)^-.$$

Thus, if $S \neq O$, $T \neq O$ and $ST = O$, then $\mathcal{N}(S)$ and $\mathcal{R}(T)^-$ are nontrivial invariant subspaces for T and S, respectively. But Problem 1.2 says that $\mathcal{N}(S)$ and $\mathcal{R}(T)^-$ are always invariant (hyperinvariant, actually) for S and T, respectively. Conclusion: If $S \neq O$, $T \neq O$ and $ST = O$, then $\mathcal{N}(S)$ and $\mathcal{R}(T)^-$ are nontrivial invariant subspaces for both for T and S.

Solution 1.4. Take any nonzero operator T in $\mathcal{B}[\mathcal{X}]$ and let S be a nontrivial solution in $\mathcal{B}[\mathcal{X}]$ to the equation

$$STS = TS.$$

If $TS = O$, then T has a nontrivial invariant subspace according to Problem 1.3. Thus suppose $TS \neq O$. Since $(I - S)TS = O$ and $S \neq I$, we get

$$\{0\} \neq \mathcal{R}(TS) \subseteq \mathcal{N}(I - S) \neq \mathcal{X}$$

so that $\mathcal{N}(I - S)$ is nontrivial. Recall: $\mathcal{N}(I - S)$ is a subspace of $\mathcal{X}$ since $(I - S)$ lies in $\mathcal{B}[\mathcal{X}]$. But $\mathcal{N}(I - S)$ is T-invariant. Indeed, if $x \in \mathcal{N}(I - S)$, then $Sx = x$ so that $Tx = TSx = STSx = STx$, and hence $Tx \in \mathcal{N}(I-S)$. Therefore, in both cases, T has a nontrivial invariant subspace.

Solution 1.5. Let T be an operator on a normed space. Note that nonzero nilpotent operators and nonscalar operators exist only on spaces with dimension greater than 1.

(a) $\mathcal{N}(T)$ is a hyperinvariant subspace for T (Problem 1.2(a)). If T is nonzero and nilpotent, then $T^n \neq O$ and $T^{n+1} = TT^n = O$ for some positive integer n so that $\mathcal{N}(T)$ is nontrivial (Problem 1.3).

(b) If T is an algebraic operator, then there exists a *minimal* polynomial p (i.e., one of minimum degree) such that $p(T) = O$. If, in addition, T is nonscalar, then the degree of p is greater than 1. Recall that a polynomial (in one complex variable, with complex coefficients) of degree $n+1$ is the product of a polynomial of degree n and a polynomial of degree 1. Thus

$$p(T) = (\lambda I - T)q(T) = O$$

for some scalar λ and some polynomial q. But $(\lambda I - T) \neq O$ because T is nonscalar and $q(T) \neq O$ because p is minimal. Therefore, according to Problems 1.2(a) and 1.3, $\mathcal{N}(\lambda I - T)$ is a nontrivial invariant subspace for every operator that commutes with $(\lambda I - T)$, and hence for every operator that commutes with T. Outcome: $\mathcal{N}(\lambda I - T)$ is a nontrivial hyperinvariant subspace for T.

Solution 1.6. Problem 1.6 is just another way to state Problem 1.2(a).

Solution 1.7. Let $\mathcal{X}$ and $\mathcal{Y}$ be normed spaces and take T in $\mathcal{B}[\mathcal{X}]$, S in $\mathcal{B}[\mathcal{Y}]$ and X in $\mathcal{B}[\mathcal{X}, \mathcal{Y}]$ such that $XT = SX$ and $\mathcal{R}(X)^- = \mathcal{Y}$. Recall that, for each x in $\mathcal{X}$, $X(\mathrm{span}\{T^n x\}_{n \geq 0}) = \mathrm{span}\{XT^n x\}_{n \geq 0}$, and $XT^n = S^n X$ for every $n \geq 0$. Thus

$$X\big(\mathrm{span}\{T^n x\}_{n \geq 0}\big) = \mathrm{span}\{S^n X x\}_{n \geq 0}$$

for each $x \in \mathcal{X}$, and therefore (since X is continuous)

$$X\big(\bigvee\{T^n x\}_{n \geq 0}\big) \subseteq \bigvee\{S^n X x\}_{n \geq 0}$$

for every $x \in \mathcal{X}$. But $X(\mathcal{X})^- = \mathcal{Y}$ and $\bigvee\{S^n X x\}_{n \geq 0}$ is closed in $\mathcal{Y}$. Hence

$$\bigvee\{T^n x\}_{n \geq 0} = \mathcal{X} \quad \text{implies} \quad \bigvee\{S^n X x\}_{n \geq 0} = \mathcal{Y}.$$

Thus if x is a cyclic vector for T, then Xx is a cyclic vector for S.

Solution 1.8. Let $\mathcal{X}$ and $\mathcal{Y}$ be normed spaces. Take T in $\mathcal{B}[\mathcal{X}]$, S in $\mathcal{B}[\mathcal{Y}]$ and X in $\mathcal{B}[\mathcal{X}, \mathcal{Y}]$ such that

$$XT = SX.$$

Let $\mathcal{M}$ be a subspace of $\mathcal{Y}$. Since $X: \mathcal{X} \to \mathcal{Y}$ is linear and continuous, the inverse image $X^{-1}(\mathcal{M})$ of $\mathcal{M}$ under X is a subspace of $\mathcal{X}$ (the inverse linear image of a linear manifold is a linear manifold, and the inverse continuous image of a closed set is closed). Now suppose $\mathcal{M}$ is invariant for S. Since $XT = SX$, $X(X^{-1}(M)) \subseteq M$ for every subset M of $\mathcal{Y}$, and $S(\mathcal{M}) \subseteq \mathcal{M}$,

$$XT(X^{-1}(\mathcal{M})) = SX(X^{-1}(\mathcal{M})) \subseteq S(\mathcal{M}) \subseteq \mathcal{M}.$$

Thus, recalling that $N \subseteq X^{-1}(X(N))$ for every subset N of $\mathcal{X}$, we get

$$T(X^{-1}(\mathcal{M})) \subseteq X^{-1}(XT(X^{-1}(\mathcal{M}))) \subseteq X^{-1}(\mathcal{M})$$

so that $X^{-1}(\mathcal{M})$ is an invariant subspace for T. Next suppose $\mathcal{M}$ is a proper subspace. If $X^{-1}(\mathcal{M}) = \mathcal{X}$, then $\mathcal{R}(X) = X(\mathcal{X}) = X(X^{-1}(\mathcal{M})) \subseteq \mathcal{M}$, and hence $\mathcal{R}(X)^- \subseteq \mathcal{M}^- = \mathcal{M} \neq \mathcal{Y}$. Thus $X^{-1}(\mathcal{M}) = \mathcal{X}$ implies $\mathcal{R}(X)^- \neq \mathcal{Y}$. Therefore,

$$\mathcal{R}(X)^- = \mathcal{Y} \quad \text{implies} \quad X^{-1}(\mathcal{M}) \neq \mathcal{X}.$$

Finally, take an arbitrary y in $\mathcal{R}(X) \cap \mathcal{M}$ so that $y = Xx \in \mathcal{M}$ for some x in $\mathcal{X}$. If $X^{-1}(\mathcal{M}) = \{0\}$ (i.e., $\{u \in \mathcal{X} : Xu \in \mathcal{M}\} = \{0\}$), then $x = 0$ and so $y = Xx = 0$. Therefore, $X^{-1}(\mathcal{M}) = \{0\}$ implies $\mathcal{R}(X) \cap \mathcal{M} = \{0\}$. Thus

$$\mathcal{R}(X) \cap \mathcal{M} \neq \{0\} \quad \text{implies} \quad X^{-1}(\mathcal{M}) \neq \{0\}.$$

Clearly, the above assumption implies $\mathcal{M} \neq \{0\}$ so that $\mathcal{M}$ is, in fact, nontrivial. Summing up: If $XT = SX$ with $\mathcal{R}(X)^- = \mathcal{Y}$, and if $\mathcal{M}$ is a nontrivial invariant subspace for S such that $\mathcal{R}(X) \cap \mathcal{M} \neq \{0\}$, then $X^{-1}(\mathcal{M})$ is a nontrivial invariant subspace for T.

Solution 1.9. Let x be an arbitrary vector in a normed space $\mathcal{X}$. Take an operator T in $\mathcal{B}[\mathcal{X}]$ and consider the set

$$\mathcal{T}_x = \{Lx \in \mathcal{X} : L \in \{T\}'\} = \bigcup_{L \in \{T\}'} Lx \subseteq \mathcal{X}.$$

If y_1 and y_2 lie in $\mathcal{T}_x$, then there exist operators L_1 and L_2 in $\{T\}'$ such that $y_1 = L_1 x$ and $y_2 = L_2 x$, and hence $y_1 + y_2 = (L_1 + L_2)x$ lies in $\mathcal{T}_x$ (because $L_1 + L_2 \in \{T\}'$). If y lies in $\mathcal{T}_x$, then there exists L in $\{T\}'$ such that $y = Lx$, and so $\alpha y = \alpha L y$ also lies in $\mathcal{T}_x$ (since $\alpha L \in \{T\}'$), for every scalar α. Outcome: as an algebra, $\{T\}'$ is a linear space, and hence $\mathcal{T}_x$ is a linear manifold of $\mathcal{X}$. Now take an arbitrary $L \in \{T\}'$. If y lies in $\mathcal{T}_x$, then $y = L_0 x$ for some L_0 in $\{T\}'$ so that $Ly = LL_0 x$ also lies in $\mathcal{T}_x$ (reason: $LL_0 \in \{T\}'$ once $\{T\}'$ is an algebra). Therefore, $L(\mathcal{T}_x) \subseteq \mathcal{T}_x$; that is, $\mathcal{T}_x$ is an invariant linear manifold for L, and hence $\mathcal{T}_x^-$ is an invariant subspace for L (Problem 1.1(b)). Since L is an arbitrary operator in $\{T\}'$, it follows that $\mathcal{T}_x^-$ is a hyperinvariant subspace for T.

Solution 1.10. Let $\mathcal{X}$ and $\mathcal{Y}$ be normed spaces. Take T in $\mathcal{B}[\mathcal{X}]$, S in $\mathcal{B}[\mathcal{Y}]$, X in $\mathcal{B}[\mathcal{X}, \mathcal{Y}]$ and Y in $\mathcal{B}[\mathcal{Y}, \mathcal{X}]$ such that

$$XT = SX \quad \text{and} \quad YS = TY.$$

If $L \in \mathcal{B}[\mathcal{X}]$ commutes with T (i.e., if $LT = TL$), then

$$(XLY)S = XLTY = XTLY = S(XLY)$$

so that XLY commutes with S. In other words, XLY lies in $\{S\}'$ for every L in $\{T\}'$. Now let $\mathcal{M}$ be a nontrivial hyperinvariant subspace for S. Thus, as $\mathcal{M}$ is hyperinvariant for S, it is invariant for XLY whenever L lies in $\{T\}'$. Take an arbitrary x in $Y(\mathcal{M})$ so that $x = Yu$ for some u in $\mathcal{M} \subseteq \mathcal{Y}$, and take an arbitrary y in $\mathcal{T}_x$ so that $y = Lx = LYu$ for some L in $\{T\}'$. Since u lies in $\mathcal{M}$ and $\mathcal{M}$ is XLY-invariant, it follows that Xy must be in $\mathcal{M}$. Therefore $X(\mathcal{T}_x) \subseteq \mathcal{M}$ and, as $\mathcal{M}$ is closed and properly included in $\mathcal{Y}$,

$$X(\mathcal{T}_x^-) \subseteq \mathcal{M}^- = \mathcal{M} \neq \mathcal{Y}$$

for every x in $Y(\mathcal{M})$ because X is continuous. Therefore, if $\mathcal{T}_x^- = \mathcal{X}$, then $\mathcal{R}(X)^- = X(\mathcal{X})^- = X(\mathcal{T}_x^-)^- \subseteq \mathcal{M} \neq \mathcal{Y}$, and hence

$$\mathcal{R}(X)^- = \mathcal{Y} \quad \text{implies} \quad \mathcal{T}_x^- \neq \mathcal{X}$$

for every x in $Y(\mathcal{M})$. But we know from Problem 1.9 that $\mathcal{T}_x^-$ is a hyperinvariant subspace for T, and we have already seen that $\mathcal{T}_x \neq \{0\}$ whenever $x \neq 0$. Thus, in order to show that $\mathcal{T}_x^-$ is a nontrivial hyperinvariant for T, we must ensure that there exists a nonzero x in $Y(\mathcal{M})$ or, equivalently, that $Y(\mathcal{M}) \neq \{0\}$. Observe that if $Y(\mathcal{M}) = \{0\}$ (i.e., if $\mathcal{M} \subseteq \mathcal{N}(Y)$), then $\mathcal{N}(Y) \cap \mathcal{M} = \mathcal{M} \neq \{0\}$, since $\mathcal{M}$ is nonzero. Thus

$$\mathcal{N}(Y) \cap \mathcal{M} = \{0\} \quad \text{implies} \quad Y(\mathcal{M}) \neq \{0\}.$$

Conclusion: If $\mathcal{R}(X)^- = \mathcal{Y}$ and $\mathcal{N}(Y) \cap \mathcal{M} = \{0\}$, then $\mathcal{T}_x^-$ is a nontrivial hyperinvariant for T for every nonzero x in $Y(\mathcal{M})$.

2

Hilbert Space Operators

Let $\mathcal{X}$ be an inner product space. Recall that $|\langle x\,;y\rangle| \leq \|x\|\,\|y\|$ (*Schwartz inequality*) and $\|x+y\|^2 = \|x\|^2 + 2\operatorname{Re}\langle x\,;y\rangle + \|y\|^2$ for every x and y in $\mathcal{X}$, where the norm $\|\ \|$ is that induced by the inner product $\langle\ ;\ \rangle$. Two vectors x and y in $\mathcal{X}$ are *orthogonal* if $\langle x\,;y\rangle = 0$. In this case we write $x \perp y$. Two subsets A and B of $\mathcal{X}$ are orthogonal (notation: $A \perp B$) if every vector in A is orthogonal to every vector in B. The *orthogonal complement* of a set A is the set $A^\perp$ made up of all vectors in $\mathcal{X}$ that are orthogonal to every vector of A. Observe that $\{0\}^\perp = \mathcal{X}$, $\mathcal{X}^\perp = \{0\}$, and $A \perp B$ if and only if $A \subseteq B^\perp$. Moreover, $A^\perp$ is a subspace (closed linear manifold) of $\mathcal{X}$, and $A \cap A^\perp \subseteq \{0\}$. In fact, $A^\perp = (A^\perp)^- = (A^-)^\perp$. If $\mathcal{M}$ is a linear manifold of $\mathcal{X}$, then $\mathcal{M} \cap \mathcal{M}^\perp = \{0\}$ and, if $\mathcal{X}$ is a Hilbert space, then $\mathcal{M}^{\perp\perp} = \mathcal{M}^-$, and $\mathcal{M}^\perp = \{0\}$ if and only if $\mathcal{M}^- = \mathcal{X}$ (see e.g., [32, §5.4]).

Take $T \in \mathcal{B}[\mathcal{X},\mathcal{Y}]$, where $\mathcal{X}$ is a Hilbert space and $\mathcal{Y}$ is an inner product space. The *adjoint* T^* of T is the unique mapping of $\mathcal{Y}$ into $\mathcal{X}$ such that

$$\langle Tx\,;y\rangle = \langle x\,;T^*y\rangle \quad \text{for every} \quad x \in \mathcal{X} \text{ and } y \in \mathcal{Y}.$$

Equivalently, $\langle T^*y\,;x\rangle = \langle y\,;Tx\rangle$ for every $x \in \mathcal{X}$ and $y \in \mathcal{Y}$ (since inner products are Hermitian symmetric). In fact, T^* is a bounded linear transformation; that is, $T^* \in \mathcal{B}[\mathcal{Y},\mathcal{X}]$. Moreover, if $\mathcal{Y}$ also is a Hilbert space, then $T^{**} = T$ and (see e.g., [32, pp. 379,380])

$$\|T^*\|^2 = \|T^*T\| = \|TT^*\| = \|T\|^2.$$

Recall that if T lies in $\mathcal{B}[\mathcal{X},\mathcal{Y}]$ and S lies in $\mathcal{B}[\mathcal{Y},\mathcal{Z}]$, where $\mathcal{X}$ and $\mathcal{Y}$ are Hilbert spaces and $\mathcal{Z}$ is an inner product space, then $(ST)^* = T^*S^*$.

Problem 2.1. If T is a bounded linear transformation of a Hilbert space $\mathcal{H}$ into a Hilbert space $\mathcal{K}$, then show that

(a) $\mathcal{N}(T) = \mathcal{R}(T^*)^\perp = \mathcal{N}(T^*T)$,

(b) $\mathcal{R}(T)^- = \mathcal{N}(T^*)^\perp = \mathcal{R}(TT^*)^-$,

(a*) $\mathcal{N}(T^*) = \mathcal{R}(T)^\perp = \mathcal{N}(TT^*)$,

(b*) $\mathcal{R}(T^*)^- = \mathcal{N}(T)^\perp = \mathcal{R}(T^*T)^-$.

An operator $T \in \mathcal{B}[\mathcal{H}]$ is *self-adjoint* (or *Hermitian*) if $T^* = T$. By the definition of the adjoint operator, $T \in \mathcal{B}[\mathcal{H}]$ is self-adjoint if and only if

$$\langle Tx\,;y \rangle = \langle x\,;Ty \rangle \quad \text{for every} \quad x,y \in \mathcal{H}.$$

It is readily verified that, if $\mathcal{H}$ is a *complex* Hilbert space, then $T \in \mathcal{B}[\mathcal{H}]$ is self-adjoint if and only if $\langle Tx\,;x \rangle \in \mathbb{R}$ for every $x \in \mathcal{H}$. This leads to a partial ordering of the set of all self-adjoint operators. Let $Q \in \mathcal{B}[\mathcal{H}]$ be a self-adjoint operator. We say that Q is *nonnegative* (notation: $Q \geq O$) if $\langle Qx\,;x \rangle \geq 0$ for every x in $\mathcal{H}$. If $\langle Qx\,;x \rangle > 0$ for every nonzero x in $\mathcal{H}$, then Q is *positive* (notation: $Q > O$). If there exists a real number $\alpha > 0$ such that $\alpha\|x\|^2 \leq \langle Qx\,;x \rangle$ for every x in $\mathcal{H}$, then Q is *strictly positive* (notation: $Q \succ O$). Trivially,

$$Q \succ O \quad \Longrightarrow \quad Q > O \quad \Longrightarrow \quad Q \geq O \quad \Longrightarrow \quad Q = Q^*.$$

If $Q \succ O$, then $\alpha\|x\| \leq \|Qx\|$ for every x in $\mathcal{H}$ (Schwarz inequality). That is, Q is *bounded below*, which is equivalent to saying that the nonnegative Q is invertible and has a bounded inverse. Conversely, if $Q \geq O$ is bounded below, then $Q \succ O$ by the next problem. Thus $Q \succ O$ *if and only if* Q *is nonnegative and invertible with a bounded inverse*. If A and B are operators on the same Hilbert space, and if $B - A \geq O$, then we write $A \leq B$. Therefore $Q \succ O$ *if and only if* $\alpha I \leq Q$ for some $\alpha > 0$. Similarly, we write $A < B$ or $A \prec B$ when $B - A > O$ or $B - A \succ O$, respectively. Observe that T^*T in $\mathcal{B}[\mathcal{H}]$ and TT^* in $\mathcal{B}[\mathcal{K}]$ are nonnegative operators for every T in $\mathcal{B}[\mathcal{H},\mathcal{K}]$ (where $\mathcal{H}$ and $\mathcal{K}$ are Hilbert spaces).

Problem 2.2. If Q is a nonnegative operator on a Hilbert space $\mathcal{H}$, then

$$|\langle Qx\,;y \rangle|^2 \leq \langle Qx\,;x \rangle \langle Qy\,;y \rangle$$

for every $x,y \in \mathcal{H}$, and hence

$$\|Qx\|^2 \leq \|Q\|\langle Qx\,;x \rangle \quad \text{for every} \quad x \in \mathcal{H}.$$

Hint: A semi-inner product space is a linear space equipped with a semi-inner product. The difference between an inner product and a semi-inner product is that a semi-inner product induces a seminorm. The difference between a norm and a seminorm is that a seminorm may also vanish for a nonzero vector.

Let $\mathcal{H}$ and $\mathcal{K}$ be Hilbert spaces and take any $T \in \mathcal{B}[\mathcal{H}, \mathcal{K}]$. Recall that

$$\|T\| = \sup_{\|x\|=\|y\|=1} |\langle Tx\,;y\rangle|,$$

and hence $T = O$ if and only if $\langle Tx\,;y\rangle = 0$ for all x in $\mathcal{H}$ and all y in $\mathcal{K}$. For self-adjoint operators we have more than that. In fact, if $T \in \mathcal{B}[\mathcal{H}]$ is self-adjoint, then (see e.g., [32, pp. 396–398])

$$\|T\| = \sup_{\|x\|=1} |\langle Tx\,;x\rangle|.$$

Thus, if T is self-adjoint, then $T = O$ if and only if $\langle Tx\,;x\rangle = 0$ for all x in $\mathcal{H}$. There are, however, operators for which this norm identity holds that are not self-adjoint. Indeed, the class of all operators T in $\mathcal{B}[\mathcal{H}]$ such that $\|T\| = \sup_{\|x\|=1} |\langle Tx\,;x\rangle|$ coincides with the class of *normaloid* operators (see e.g., [32, p. 465]). Normaloid will be considered in Problem 7.9.

Problem 2.3. Let A and Q be operators on a Hilbert space $\mathcal{H}$. Prove the following assertions.

(a) $-I \le A^* = A \le I$ if and only if $A^* = A$ and $\|A\| \le 1$.

(b) $O < Q < I \iff O \le Q$ and $\|Q\| \le 1 \iff Q^* = Q$ and $Q^2 \le Q$.

An operator on a Hilbert space is *normal* if it commutes with its adjoint. That is, $T \in \mathcal{B}[\mathcal{H}]$ is normal if $T^*T = TT^*$. It is clear that every self-adjoint operator is normal.

Problem 2.4. Let T be an operator on a Hilbert space $\mathcal{H}$. Show that the following assertions are pairwise equivalent.

(a) T is normal.

(b) $\|T^*x\| = \|Tx\|$ for every $x \in \mathcal{H}$.

(c) T^n is normal for every integer $n \ge 1$.

(d) $\|T^{*n}x\| = \|T^n x\|$ for every $x \in \mathcal{H}$ and every integer $n \ge 1$.

An *isometry* between metric spaces is a map that preserves distance. In a normed space setting we shall be concerned with *linear* isometries. Thus, from now on, "isometry" will mean "linear isometry".

Problem 2.5. Let V be a linear transformation of normed space $\mathcal{X}$ into a normed space $\mathcal{Y}$. Show that the assertions below are equivalent.

(a) V is an isometry   (i.e., $\|Vx - Vy\| = \|x - y\|$ for every $x, y \in \mathcal{X}$).

(b) $\|Vx\| = \|x\|$ for every $x \in \mathcal{X}$ (thus $\|V\| = 1$ if $\mathcal{X} \neq \{0\}$).

If $\mathcal{Y} = \mathcal{X}$, then each of the above assertions also is equivalent to

(c) $\|V^n x\| = \|x\|$ for every $x \in \mathcal{X}$ and every integer $n \geq 1$.

Now suppose $\mathcal{X}$ and $\mathcal{Y}$ are inner product spaces. Show that V is an isometry if and only if it preserves inner product; that is, if and only if

(d) $\langle Vx ; Vy \rangle = \langle x ; y \rangle$ for every $x, y \in \mathcal{X}$.

Moreover, if $\mathcal{X}$ is a Hilbert space, then V is an isometry if and only if

(e) $V^*V = I$, where I is the identity on $\mathcal{X}$.

If, in addition, $\mathcal{X} = \mathcal{Y}$, then show that V is an isometry if and only if

(f) $V^{*n}V^n = I$ for every integer $n \geq 1$.

Assertion (b) in the above problem (i.e., $\|Vx\| = \|x\|$ for every $x \in \mathcal{X}$) ensures that the kernel of an isometry is null. That is, if $V \in \mathcal{B}[\mathcal{X}, \mathcal{Y}]$ is an isometry, then $\mathcal{N}(V) = \{0\}$. Since V is linear, $\mathcal{N}(V) = \{0\}$ if and only if V is injective. Outcome: every isometry is injective. Hence a surjective isometry is invertible, which means an isometric isomorphism (an isomorphism between linear spaces is an invertible linear transformation). A surjective isometry between inner product spaces is a *unitary transformation* (i.e., a unitary transformation is precisely an invertible isometry between inner product spaces). In other words, according to assertion (d) in the above problem, a unitary transformation of an inner product space *onto* an inner product space is an isomorphism that preserves inner product. Two inner product spaces are *unitarily equivalent* if there exists a unitary transformation between them. Recall that two operators on normed spaces are similar if they are intertwined by an invertible bounded linear transformation with a bounded inverse. Two operators on inner product spaces are *unitarily equivalent* if they are intertwined by a unitary transformation. That is, T in $\mathcal{B}[\mathcal{X}]$ and S in $\mathcal{B}[\mathcal{Y}]$ are unitarily equivalent if $UT = SU$ for some unitary transformation U in $\mathcal{B}[\mathcal{X}, \mathcal{Y}]$. Note that the inverse of an invertible isometry is again an isometry. Thus U^{-1} lies in $\mathcal{B}[\mathcal{Y}, \mathcal{X}]$ and is unitary too.

Now let $\mathcal{H}$ and $\mathcal{K}$ be Hilbert spaces. A *coisometry* is a transformation T in $\mathcal{B}[\mathcal{H}, \mathcal{K}]$ such that its adjoint T^* in $\mathcal{B}[\mathcal{K}, \mathcal{H}]$ is an isometry. Then assertion (e) of the previous problem says that T is a coisometry if and only if $TT^* = I$, where I is the identity on $\mathcal{K}$.

Problem 2.6. Take $U \in \mathcal{B}[\mathcal{H}, \mathcal{K}]$, where $\mathcal{H}$ and $\mathcal{K}$ are Hilbert spaces. Show that the following assertions are pairwise equivalent.

(a) U is unitary   (i.e., U is a surjective isometry).

(b) U is invertible and $U^{-1} = U^*$.

(c) U is invertible and $\|U\| = \|U^{-1}\| = 1$.

(d) U is invertible, $\|U\| \leq 1$ and $\|U^{-1}\| \leq 1$.

(e) $U^*U = I$ (identity on $\mathcal{H}$) and $UU^* = I$ (identity on $\mathcal{K}$).

(f) U is an isometry and a coisometry.

(g) $\|Ux\| = \|x\|$ and $\|U^*y\| = \|y\|$ for every $x \in \mathcal{H}$ and every $y \in \mathcal{K}$.

If $\mathcal{H} = \mathcal{K}$, then each of the above is equivalent to each of the following.

(h) U is a normal isometry.

(i) $\|U^*x\| = \|Ux\| = \|x\|$ for every $x \in \mathcal{H}$.

(j) $\|U^{*n}x\| = \|U^n x\| = \|x\|$ for every $x \in \mathcal{H}$ and every integer $n \geq 1$.

(k) $U^{*n}U^n = U^n U^{*n} = I$ every integer $n \geq 1$.

Let $\mathcal{M}$ and $\mathcal{N}$ be linear manifolds of a linear space $\mathcal{X}$. They are *algebraic complements* of each other if $\mathcal{M} + \mathcal{N} = \mathcal{X}$ and $\mathcal{M} \cap \mathcal{N} = \{0\}$. In this case, each vector x in $\mathcal{X}$ can be uniquely written as $x = u + v$ with u in $\mathcal{M}$ and v in $\mathcal{N}$. A *projection* is an idempotent linear transformation of a linear space into itself. That is, if $\mathcal{X}$ is a linear space, then $P \colon \mathcal{X} \to \mathcal{X}$ is a projection if it is linear and $P = P^2$. If a projection P is such that $O \neq P \neq I$, then it is a *nontrivial projection*. It is easy to verify that the range of a projection is the set of all its fixed points: if $P = P^2$, then $\mathcal{R}(P) = \{x \in \mathcal{X} \colon Px = x\}$. Moreover, range and kernel of a projection are algebraic complements,

$$\mathcal{R}(P) + \mathcal{N}(P) = \mathcal{X} \quad \text{and} \quad \mathcal{R}(P) \cap \mathcal{N}(P) = \{0\},$$

and $I - P$ is a projection as well, the *complementary projection* of P, where

$$\mathcal{R}(I - P) = \mathcal{N}(P) \quad \text{and} \quad \mathcal{N}(I - P) = \mathcal{R}(P)$$

(see e.g., [32, §2.9]). Recall that subspace means closed linear manifold. If a pair of subspaces of a normed space are algebraic complements, then they are called *complementary subspaces*. An *orthogonal projection* on an inner product space $\mathcal{X}$ is a projection $P \colon \mathcal{X} \to \mathcal{X}$ such that $\mathcal{R}(P) \perp \mathcal{N}(P)$. Therefore, if P is an orthogonal projection, then so is the complementary projection $I - P$. Orthogonal projections are continuous (i.e., $P \in \mathcal{B}[\mathcal{X}]$). Indeed, since $\mathcal{R}(P)$ and $\mathcal{N}(P)$ are algebraic complements, every x in $\mathcal{X}$ can be written as $x = u + v$ with u in $\mathcal{R}(P)$ and v in $\mathcal{N}(P)$, and hence (since P is linear, $Pu = u$, $Pv = 0$ and $u \perp v$) by the Pythagorean Theorem we get $\|Px\|^2 = \|Pu + Pv\|^2 = \|u\|^2 \leq \|u\|^2 + \|v\|^2 = \|x\|^2$ so that $\|P\| \leq 1$. Since $P \in \mathcal{B}[\mathcal{X}]$, the kernels $\mathcal{N}(P)$ and $\mathcal{N}(I - P)$ are closed in $\mathcal{X}$. Then $\mathcal{R}(P)$ and $\mathcal{N}(P)$ are orthogonal complementary subspaces of $\mathcal{X}$, and therefore $\mathcal{N}(P) = \mathcal{R}(P)^{\perp}$ and $\mathcal{R}(P) = \mathcal{N}(P)^{\perp}$.

Problem 2.7. If $P \in \mathcal{B}[\mathcal{H}]$ is a nonzero projection on a Hilbert space $\mathcal{H}$, then show that the following assertions are pairwise equivalent.

(a) P is an orthogonal projection.

(b) P is nonnegative.

(c) P is self-adjoint.

(d) P is normal.

(e) $\|P\| = 1$.

(f) $\|P\| \leq 1$.

Problem 2.8. Let $P_1 \in \mathcal{B}[\mathcal{X}]$ and $P_2 \in \mathcal{B}[\mathcal{X}]$ be orthogonal projections on an inner product space $\mathcal{X}$. Show that the following assertions are pairwise equivalent.

(a) $\mathcal{R}(P_1) \perp \mathcal{R}(P_2)$.

(b) $P_1 P_2 = O$.

(c) $P_2 P_1 = O$.

(d) $\mathcal{R}(P_2) \subseteq \mathcal{N}(P_1)$.

(e) $\mathcal{R}(P_1) \subseteq \mathcal{N}(P_2)$.

If two orthogonal projections P_1 and P_2 on an inner product space $\mathcal{X}$ satisfy any of the above equivalent assertions, then they are said to be *orthogonal to each other* (or *mutually orthogonal*). The next problem requires that the orthogonal projections act on a Hilbert space. (Why?)

Problem 2.9. Let $P_1 \in \mathcal{B}[\mathcal{H}]$ and $P_2 \in \mathcal{B}[\mathcal{H}]$ be orthogonal projections on a Hilbert space $\mathcal{H}$. Show that the assertions below are pairwise equivalent.

(a) $P_1 \leq P_2$.

(b) $\|P_1 x\| \leq \|P_2 x\|$ for every $x \in \mathcal{H}$.

(c) $\mathcal{N}(P_2) \subseteq \mathcal{N}(P_1)$.

(d) $\mathcal{R}(P_1) \subseteq \mathcal{R}(P_2)$.

(e) $P_2 P_1 = P_1$.

(f) $P_1 P_2 = P_1$.

(g) $P_2 P_1 P_2 = P_1$.

(h) $P_2 - P_1$ is an orthogonal projection.

Solutions

Solution 2.1. Take any T in $\mathcal{B}[\mathcal{H}, \mathcal{K}]$, where $\mathcal{H}$ and $\mathcal{K}$ are Hilbert spaces, and consider the assertions (a), (b), (a*) and (b*) in the problem statement. Recall that $x \in \mathcal{R}(T^*)^\perp$ if and only if $\langle x ; T^* y \rangle = 0$ for every y in $\mathcal{K}$. The definition of adjoint says that this is equivalent to $\langle Tx ; y \rangle = 0$ for every y in $\mathcal{K}$, which means $Tx = 0$; that is, $x \in \mathcal{N}(T)$. Hence

$$\mathcal{R}(T^*)^\perp = \mathcal{N}(T).$$

Moreover, since $\|Tx\|^2 = \langle Tx ; Tx \rangle = \langle T^*Tx ; x \rangle$ for every x in $\mathcal{H}$, it follows that $\mathcal{N}(T^*T) \subseteq \mathcal{N}(T)$. But $\mathcal{N}(T) \subseteq \mathcal{N}(T^*T)$ trivially, and so

$$\mathcal{N}(T) = \mathcal{N}(T^*T),$$

completing the proof of (a). Since (a) holds for every T in $\mathcal{B}[\mathcal{H}, \mathcal{K}]$, it also holds for T^* in $\mathcal{B}[\mathcal{K}, \mathcal{H}]$ and for TT^* in $\mathcal{B}[\mathcal{K}]$. Thus (recall: $\mathcal{M}^{\perp\perp} = \mathcal{M}^-$ for every linear manifold $\mathcal{M}$ of any Hilbert space, and $T^{**} = T$),

$$\mathcal{R}(T)^- = \mathcal{R}(T^{**})^{\perp\perp} = \mathcal{N}(T^*)^\perp = \mathcal{N}(T^{**}T^*)^\perp$$
$$= \mathcal{N}(TT^*)^\perp = \mathcal{R}((TT^*)^*)^{\perp\perp} = \mathcal{R}(T^{**}T^*)^{\perp\perp} = \mathcal{R}(TT^*)^-,$$

which proves (b). Since $T^{**} = T$ we get the dual expressions (a*) and (b*).

Solution 2.2. Take a nonnegative operator $Q \in \mathcal{B}[\mathcal{H}]$ on a Hilbert space $\mathcal{H}$ and consider the function $\langle \ ; \ \rangle_Q : \mathcal{H} \times \mathcal{H} \to \mathbb{F}$ given by

$$\langle x ; y \rangle_Q = \langle Qx ; y \rangle$$

for every $x, y \in \mathcal{H}$. It is easy to verify that the function $\langle \ ; \ \rangle_Q$ defines a semi-inner product on $\mathcal{H}$. Let $\| \ \|_Q$ be the seminorm induced on $\mathcal{H}$ by the semi-inner product $\langle \ ; \ \rangle_Q$ so that $\|x\|_Q = \langle Qx ; x \rangle^{\frac{1}{2}}$ for every $x \in \mathcal{H}$. Since the Schwarz inequality holds in a semi-inner product space, it follows that

$$|\langle Qx ; y \rangle|^2 = |\langle x ; y \rangle_Q|^2 \le \|x\|_Q^2 \|y\|_Q^2 = \langle Qx ; x \rangle \langle Qy ; y \rangle$$

for every $x, y \in \mathcal{H}$. In particular, by setting $y = Qx$ we get

$$\|Qx\|^4 = \langle Qx ; Qx \rangle^2 \le \langle Qx ; x \rangle \langle Q^2 x ; Qx \rangle$$
$$\le \langle Qx ; x \rangle \|Q^2 x\| \|Qx\| \le \langle Qx ; x \rangle \|Q\| \|Qx\|^2,$$

which implies that $\|Qx\|^2 \le \|Q\| \langle Qx ; x \rangle$, for every $x \in \mathcal{H}$.

Solution 2.3. Suppose A is a self-adjoint operator on a Hilbert space $\mathcal{H}$. Thus $\langle Ax ; x \rangle \in \mathbb{R}$ of all $x \in \mathcal{H}$, and so

$$\langle (\pm A - I)x ; x \rangle = \pm \langle Ax ; x \rangle - \|x\|^2 \le |\langle Ax ; x \rangle| - \|x\|^2 \le (\|Ax\| - \|x\|)\|x\|$$

for every $x \in \mathcal{H}$ by the Schwarz inequality.

(a) If $\|Ax\| \leq \|x\|$ for every x in $\mathcal{H}$, then $(\pm A - I) \leq O$; that is, $\|A\| \leq 1$ implies $-I \leq A \leq I$. On the other hand, if $(\pm A - I) \leq O$, then $\pm \langle Ax; x \rangle \leq \|x\|^2$ for every x in $\mathcal{H}$ so that $\sup_{\|x\|=1} |\langle Ax; x \rangle| \leq 1$. But $\|A\| = \sup_{\|x\|=1} |\langle Ax; x \rangle|$ because A is self-adjoint, and therefore $-I \leq A \leq I$ implies $\|A\| \leq 1$.

(b) Assertion (a) ensures: $O \leq Q \leq I$ if and only if $O \leq Q$ and $\|Q\| \leq 1$. If Q is a nonnegative contraction (that is, if $O \leq Q$ and $\|Q\| \leq 1$), then we get $\langle Q^2 x; x \rangle = \langle Qx; Qx \rangle = \|Qx\|^2 \leq \langle Qx; x \rangle$ for every x in $\mathcal{H}$ by Problem 2.2 so that $Q^2 \leq Q$. Conversely, if Q is a self-adjoint operator and $Q^2 \leq Q$, then $\|Qx\|^2 = \langle Q^2 x; x \rangle \leq \langle Qx; x \rangle \leq \|Qx\| \|x\|$, and hence $\|Qx\| \leq \|x\|$ for every x in $\mathcal{H}$; that is, $\|Q\| \leq 1$. Thus $O \leq Q$ and $\|Q\| \leq 1$ if and only if $Q^* = Q$ and $Q^2 \leq Q$.

Solution 2.4. Take an arbitrary operator $T \in \mathcal{B}[\mathcal{H}]$ on a Hilbert space $\mathcal{H}$. Since $\|T^*x\|^2 - \|Tx\|^2 = \langle T^*x; T^*x \rangle - \langle Tx; Tx \rangle = \langle (TT^* - T^*T)x; x \rangle$ for every $x \in \mathcal{H}$, and since $TT^* - T^*T$ is self-adjoint, it follows that T is normal if and only if $\|T^*x\| = \|Tx\|$ for every $x \in \mathcal{H}$. Since $T^{*n} = T^{n*}$ for each $n \geq 1$, the above equivalence ensures that T^n is normal for every $n \geq 1$ if and only if $\|T^{*n}x\| = \|T^n x\|$ for every $x \in \mathcal{H}$ and every $n \geq 1$. Finally, it follows by induction that T commutes with T^* if and only if T^n commutes with T^{*n} for every $n \geq 1$. Recalling again that $T^{*n} = T^{n*}$ for each $n \geq 1$, this implies that T is normal if and only if T^n is normal every $n \geq 1$.

Solution 2.5. Let $\mathcal{X}$ and $\mathcal{Y}$ be normed spaces and take a *linear* transformation $V: \mathcal{X} \to \mathcal{Y}$. Note that, since V is linear,

(a) $\|Vx - Vy\| = \|x - y\|$ for every $x, y \in \mathcal{X}$

is trivially equivalent to

(b) $\|Vx\| = \|x\|$ for every $x \in \mathcal{X}$.

Set $y = 0$ in (a) to get (b), and $x = u - v$ in (b) to get (a). Suppose $\mathcal{X} = \mathcal{Y}$ so that V is a map of $\mathcal{X}$ into itself, and hence we may consider the positive integral powers V^n of it. If

(c) $\|V^n x\| = \|x\|$ for every $x \in \mathcal{X}$ and every integer $n \geq 1$,

then (b) holds tautologically. Conversely, (b) ensures (c) for $n = 1$. If (c) holds for some integer $n \geq 1$, then $\|V^{n+1}x\| = \|V^n Vx\| = \|Vx\| = \|x\|$ for every $x \in \mathcal{X}$ whenever (b) holds, and hence (b) implies (c) by induction. Now suppose $\mathcal{X}$ and $\mathcal{Y}$ are inner product spaces. If

(d) $\langle Vx; Vy \rangle = \langle x; y \rangle$ for every $x, y \in \mathcal{X}$,

then $\|Vx\|^2 = \langle Vx; Vx \rangle = \langle x; x \rangle = \|x\|^2$ for all x in $\mathcal{X}$ so that (d) implies (b). Since V is linear, (b) implies (d) by the polarization identity. Indeed, the *polarization identity* says that

$$\langle x\,;y\rangle \;=\; \tfrac{1}{4}\big(\|x+y\|^2 - \|x-y\|^2 + i\|x+iy\|^2 - i\|x-iy\|^2\big),$$

or

$$\langle x\,;y\rangle \;=\; \tfrac{1}{4}\big(\|x+y\|^2 - \|x-y\|^2\big),$$

for every $x,y \in \mathcal{X}$, whether $\mathcal{X}$ is a complex or a real inner product space, respectively. Finally, if (d) holds, then $V \in \mathcal{B}[\mathcal{X},\mathcal{Y}]$ (because (b) holds). Therefore, if $\mathcal{X}$ is a Hilbert space, then consider the adjoint $V^* \in \mathcal{B}[\mathcal{Y},\mathcal{X}]$ of V so that $\langle(V^*V - I)x\,;y\rangle = \langle V^*Vx\,;y\rangle - \langle x\,;y\rangle = \langle Vx\,;Vy\rangle - \langle x\,;y\rangle$, and hence (d) holds if and only if $\langle(V^*V - I)x\,;y\rangle = 0$ for every $x,y \in \mathcal{X}$, which means that $(V^*V - I)x = 0$ for every $x \in \mathcal{X}$, where I is the identity on $\mathcal{X}$. That is, (d) holds if and only if

(e) $V^*V = I$.

If $\mathcal{X} = \mathcal{Y}$, then a trivial induction shows that (e) is equivalent to

(f) $V^{n*}V^n = I$ for every integer $n \geq 1$.

Solution 2.6. Let U be a transformation in $\mathcal{B}[\mathcal{H},\mathcal{K}]$. If U is unitary, then it is an invertible isometry, and hence there exists $U^{-1}\colon \mathcal{K} \to \mathcal{H}$ such that $UU^{-1} = I$ (where I is the identity on $\mathcal{K}$). According to Problem 2.5(d) the isometry U preserves inner product; that is, $\langle x_1\,;x_2\rangle = \langle Ux_1\,;Ux_2\rangle$ for every $x_1, x_2 \in \mathcal{H}$. Hence

$$\langle x\,;U^{-1}y\rangle \;=\; \langle Ux\,;UU^{-1}y\rangle \;=\; \langle Ux\,;y\rangle \;=\; \langle x\,;U^*y\rangle$$

for every $x \in \mathcal{H}$ and $y \in \mathcal{K}$. Therefore, since the adjoint U^* of U is unique, it follows that $U^{-1} - U^* \in \mathcal{B}[\mathcal{K},\mathcal{H}]$. Conversely, if U is invertible and if $U^{-1} = U^*$, then $U^*U = I$ (now I denotes the identity on $\mathcal{H}$), and hence U is an isometry according to Problem 2.5(e). Thus the assertions

(a) U is unitary,

(b) U is invertible and $U^{-1} = U^*$,

are equivalent. But $1 = \|U^{-1}U\| = \|U^*U\| = \|U\|^2 = \|U^*\|^2 = \|U^{-1}\|^2$ if (b) holds so that (b) implies (c), which implies (d) trivially, and (d) implies (a); if (d) holds, then $\|x\| = \|U^{-1}Ux\| \leq \|Ux\| \leq \|x\|$ for each $x \in \mathcal{H}$. The remaining equivalences follow at once by (a), (b) and Problems 2.4 and 2.5.

Solution 2.7. Let P in $\mathcal{B}[\mathcal{H}]$ be a nonzero projection on a Hilbert space $\mathcal{H}$. Recall that $\mathcal{R}(P) \perp \mathcal{N}(P)$ implies $\mathcal{R}(P) = \mathcal{N}(P)^\perp$. But Problem 2.1 says that $\mathcal{N}(P)^\perp = \mathcal{R}(P^*)^-$. Hence $\mathcal{R}(P) = \mathcal{R}(P^*)^-$. Therefore, if x is an arbitrary vector in $\mathcal{H}$, then $Px = \lim P^*x_n$ for some sequence $\{x_n\}$ of vectors in $\mathcal{H}$ so that (since P^* is bounded; that is, since it is continuous) $P^*Px = \lim P^{*2}x_n = \lim P^{2*}x_n = \lim P^*x_n = Px$ (because $P = P^2$). Thus $P^*P = P$, and hence P is nonnegative (reason: this shows that P is self-adjoint and we know that P^*P is always nonnegative). Outcome:

$$\mathcal{R}(P) \perp \mathcal{N}(P) \quad\Longrightarrow\quad P \geq O \quad\Longrightarrow\quad P^* = P \quad\Longrightarrow\quad P^*P = PP^*.$$

Conversely, if P is normal, then $\|P^*x\| = \|Px\|$ for every x in $\mathcal{H}$ (Problem 2.4) so that $\mathcal{N}(P^*) = \mathcal{N}(P)$. But $\mathcal{R}(P) = \mathcal{R}(P)^-$ (the range of a *bounded* projection P is closed because it coincides with the kernel of $I - P$), and $\mathcal{R}(P)^- = \mathcal{N}(P^*)^\perp$ by Problem 2.1. Therefore, $\mathcal{R}(P) = \mathcal{N}(P)^\perp$ so that $\mathcal{R}(P) \perp \mathcal{N}(P)$. That is,

$$P^*P = PP^* \quad \Longrightarrow \quad \mathcal{R}(P) \perp \mathcal{N}(P).$$

Recall that $\|T\|^2 = \|T^*T\|$ for every Hilbert space operator T. If P is an orthogonal projection, then it is self-adjoint, and hence $P^* = P = P^2 \neq O$ if it is nonzero. Thus $\|P\|^2 = \|P^*P\| = \|P^2\| = \|P\| \neq 0$ so that

$$\mathcal{R}(P) \perp \mathcal{N}(P) \quad \Longrightarrow \quad \|P\| = 1 \quad \Longrightarrow \quad \|P\| \leq 1.$$

Conversely, take an arbitrary u in $\mathcal{N}(P)^\perp$. Since P is a projection, it follows that $\mathcal{R}(I-P) = \mathcal{N}(P)$, and hence $(I-P)u$ lies in $\mathcal{N}(P)$. Then $(I - P)u \perp u$ so that $0 = \langle (I - P)u \, ; u \rangle = \|u\|^2 - \langle Pu ; u \rangle$. If $\|P\| \leq 1$, then we get $\|u\|^2 = \langle Pu ; u \rangle \leq \|Pu\| \|u\| \leq \|u\|^2$, and hence $\|Pu\| = \|u\| = \langle Pu ; u \rangle^{\frac{1}{2}}$. Therefore, $\|(I - P)u\|^2 = \|Pu - u\|^2 = \|Pu\|^2 - 2\operatorname{Re}\langle Pu ; u \rangle + \|u\|^2 = 0$, and so $u \in \mathcal{N}(I - P) = \mathcal{R}(P)$. Thus

$$\|P\| \leq 1 \quad \Longrightarrow \quad \mathcal{N}(P)^\perp \subseteq \mathcal{R}(P).$$

If $\mathcal{N}(P)^\perp \subseteq \mathcal{R}(P)^-$, then $\mathcal{R}(P^*) \subseteq \mathcal{N}(P^*)^\perp$, according to Problem 2.1, and so $\mathcal{R}(P^*) \perp \mathcal{N}(P^*)$. But P^* is a projection whenever P is ($P = P^2$ implies $P^* = P^{2*} = P^{*2}$). Thus P^* is an orthogonal projection so that P^* is self-adjoint. Hence $P = P^{**} = P^*$ is an orthogonal projection. Then

$$\mathcal{N}(P)^\perp \subseteq \mathcal{R}(P) \quad \Longrightarrow \quad \mathcal{R}(P) \perp \mathcal{N}(P).$$

Solution 2.8. Consider the assertions (a)–(e) in the problem statement. Recall that $\mathcal{R}(P_1) \perp \mathcal{R}(P_2)$ if and only if $\mathcal{R}(P_2) \subseteq \mathcal{R}(P_1)^\perp$, and also that $\mathcal{R}(P_1)^\perp = \mathcal{N}(P_1)$. Therefore, (a)$\Leftrightarrow$(d). Similarly we get (a)$\Leftrightarrow$(e) (reason: $\mathcal{R}(P_1) \perp \mathcal{R}(P_2)$ if and only if $\mathcal{R}(P_2) \perp \mathcal{R}(P_1)$). Since $\mathcal{R}(P_2) \subseteq \mathcal{N}(P_1)$ if and only if $P_1 P_2 = O$, it follows (d)$\Leftrightarrow$(b). Swap P_1 and P_2 to get (e)$\Leftrightarrow$(c).

Solution 2.9. Consider the assertions (a)–(h) in the problem statement. If P is an orthogonal projection, then $\langle Px ; x \rangle = \|Px\|^2$ for every x in $\mathcal{H}$, so that (a)$\Rightarrow$(b). Clearly, (b)$\Rightarrow$(c). Since $A \subseteq B$ implies $B^\perp \subseteq A^\perp$, and since $\mathcal{R}(P) = \mathcal{N}(P)^\perp$, we get $\mathcal{R}(P_1) = \mathcal{N}(P_1)^\perp \subseteq \mathcal{N}(P_2)^\perp = \mathcal{R}(P_2)$ whenever (c) holds, and hence (c)$\Rightarrow$(d). Recall: the range of an idempotent operator is the set of all its fixed points. Thus (d) implies $P_2 P_1 x = P_1 x$ for every x in $\mathcal{H}$, and therefore (d)$\Rightarrow$(e). Orthogonal projections are self-adjoint so that (e)$\Leftrightarrow$(f). Since (f) implies (e), we get (f)$\Rightarrow$(g) and $(P_2 - P_1)^2 = P_2 - P_1$ if (f) holds. Thus (f)$\Rightarrow$(h) because $P_2 - P_1$ is self-adjoint. Orthogonal projections are nonnegative so that (h)$\Rightarrow$(a). Finally, it is clear that (g)$\Rightarrow$(d).

3
Convergence and Stability

Let $\mathcal{X}$ and $\mathcal{Y}$ be normed spaces and let $\{T_n\}$ be a $\mathcal{B}[\mathcal{X}, \mathcal{Y}]$-valued sequence (i.e., a sequence of transformations in $\mathcal{B}[\mathcal{X}, \mathcal{Y}]$). If $\{T_n\}$ converges in the normed space $\mathcal{B}[\mathcal{X}, \mathcal{Y}]$; that is, if there exists T in $\mathcal{B}[\mathcal{X}, \mathcal{Y}]$ such that

$$\|T_n - T\| \to 0,$$

then we say that $\{T_n\}$ *converges uniformly* to T. This (unique) $T \in \mathcal{B}[\mathcal{X}, \mathcal{Y}]$ is called the *uniform limit* of $\{T_n\}$. Notation: $T_n \overset{u}{\longrightarrow} T$. If $\{T_n\}$ does not converge uniformly to T, then we write $T_n \overset{u}{\nrightarrow} T$. The $\mathcal{Y}$-valued sequence $\{T_n x\}$ converges in $\mathcal{Y}$ for every x in $\mathcal{X}$ if and only if there exists a (unique) linear transformation T of $\mathcal{X}$ into $\mathcal{Y}$ such that

$$\|(T_n - T)x\| \to 0 \quad \text{for every} \quad x \in \mathcal{X}.$$

Moreover, if $\mathcal{X}$ is a Banach space, then $T \in \mathcal{B}[\mathcal{X}, \mathcal{Y}]$ (see e.g., [32, p. 243]). If the above convergence holds for some T in $\mathcal{B}[\mathcal{X}, \mathcal{Y}]$, then we say that $\{T_n\}$ *converges strongly* to T, which is called the *strong limit* of $\{T_n\}$. Notation: $T_n \overset{s}{\longrightarrow} T$. If $\{T_n\}$ does not converge strongly to T, then we write $T_n \overset{s}{\nrightarrow} T$. Since $0 \le \|(T_n - T)x\| \le \|T_n - T\| \|x\|$ for every x in $\mathcal{X}$ whenever T lies in $\mathcal{B}[\mathcal{X}, \mathcal{Y}]$, it follows that uniform convergence implies strong convergence:

$$T_n \overset{u}{\longrightarrow} T \quad \Longrightarrow \quad T_n \overset{s}{\longrightarrow} T.$$

An operator T on a normed space $\mathcal{X}$ is *uniformly stable* if the power sequence $\{T^n\}$ converges uniformly to the null operator; that is, if $T^n \overset{u}{\longrightarrow} O$ (equivalently, if $\|T^n\| \to 0$). It is *strongly stable* if $\{T^n\}$ converges strongly to the null operator; that is, if $T^n \xrightarrow{s} O$ (equivalently, if $\|T^n x\| \to 0$ for every x in $\mathcal{X}$). Clearly, uniform stability implies strong stability.

Take any $p \geq 1$ and consider the Banach space ℓ_+^p of all p-summable scalar sequences equipped with its usual norm. Now let $a = \{\alpha_k\}_{k=0}^\infty \in \ell_+^\infty$ be an arbitrary bounded scalar sequence so that

$$\|a\|_\infty = \sup_k |\alpha_k| < \infty,$$

and consider the *diagonal* mapping $D: \ell_+^p \to \ell_+^p$ defined by

$$Dx = \{\alpha_k \xi_k\}_{k=0}^\infty \quad \text{for every} \quad x = \{\xi_k\}_{k=0}^\infty \in \ell_+^p$$

(i.e., $D(\xi_0, \xi_1, \xi_2, ...) = (\alpha_0 \xi_0, \alpha_1 \xi_1, \alpha_2 \xi_2, ...)$). Common notation includes the usual representation of D as an infinite diagonal matrix:

$$D = \text{diag}(\{\alpha_k\}_{k=0}^\infty) = \text{diag}(\alpha_0, \alpha_1, \alpha_2, ...) = \begin{pmatrix} \alpha_0 & & & \\ & \alpha_1 & & \\ & & \alpha_2 & \\ & & & \ddots \end{pmatrix}.$$

This indeed is a bounded linear transformation (i.e., $D \in \mathcal{B}[\ell_+^p]$), referred to as the *diagonal operator* on ℓ_+^p. Linearity is trivially verified and boundedness will be checked below. It is also readily verified that, on the Hilbert space ℓ_+^2, the adjoint of D is the diagonal $D^* = \text{diag}(\{\overline{\alpha_k}\}_{k=0}^\infty)$.

Problem 3.1. Take a bounded sequence $a = \{\alpha_k\}_{k=0}^\infty \in \ell_+^\infty$ and consider the (linear) diagonal mapping $D = \text{diag}(\{\alpha_k\}_{k=0}^\infty): \ell_+^p \to \ell_+^p$. Show that

(a) D is bounded and $\|D\| = \|a\|_\infty$,

(b) D is uniformly stable if and only if $\|D\| < 1$ (i.e., $\sup_k |\alpha_k| < 1$),

(c) D is strongly stable if and only if $|\alpha_k| < 1$ for every $k \geq 0$.

(d) Now exhibit a strongly stable diagonal that is not uniformly stable.

Let $\mathcal{X}$ and $\mathcal{Y}$ be normed spaces and let $\mathcal{Y}^*$ be the dual space of $\mathcal{Y}$. That is, $\mathcal{Y}^* = \mathcal{B}[\mathcal{Y}, \mathbb{F}]$ is the Banach space of all bounded linear functionals on $\mathcal{Y}$, where $\mathbb{F}$ stands for the scalar field; either $\mathbb{F} = \mathbb{R}$ or $\mathbb{F} = \mathbb{C}$. Let $\{T_n\}$ be a $\mathcal{B}[\mathcal{X}, \mathcal{Y}]$-valued sequence. The scalar-valued sequence $\{f(T_n x)\}$ converges in $\mathbb{F}$ for every x in $\mathcal{X}$ and every f in $\mathcal{Y}^*$ if and only if there exists a (unique) linear transformation T of $\mathcal{X}$ into $\mathcal{Y}$ such that $f((T_n - T)x) \to 0$ every x in $\mathcal{X}$ and every f in $\mathcal{Y}^*$. Equivalently, such that

$$f(T_n x) \to f(Tx) \quad \text{for every} \quad x \in \mathcal{X} \quad \text{and every} \quad f \in \mathcal{Y}^*.$$

Moreover, if $\mathcal{X}$ is a Banach space, then $T \in \mathcal{B}[\mathcal{X}, \mathcal{Y}]$ (see e.g., [32, p. 307]). If the above convergence holds for some T in $\mathcal{B}[\mathcal{X}, \mathcal{Y}]$, then we say that $\{T_n\}$

converges weakly to T, which is called the *weak limit* of $\{T_n\}$. Notation: $T_n \xrightarrow{w} T$. If $\{T_n\}$ does not converge weakly to T, then we write $T_n \not\xrightarrow{w} T$. Since $0 \leq |f((T_n - T)x)| \leq \|f\|\|(T_n - T)x\|$ for every x in $\mathcal{X}$ and every f in $\mathcal{Y}^*$, it follows that strong convergence implies weak convergence:

$$T_n \xrightarrow{s} T \quad \Longrightarrow \quad T_n \xrightarrow{w} T.$$

If $\mathcal{X}$ is a Banach space, then the Banach–Steinhaus Theorem ensures that

$$T_n \xrightarrow{w} T \quad \Longrightarrow \quad \sup_n \|T_n\| < \infty.$$

Problem 3.2. Let $\{T_n\}$ be a sequence in $\mathcal{B}[\mathcal{X}, \mathcal{Y}]$ and $\{S_n\}$ be a sequence in $\mathcal{B}[\mathcal{Y}, \mathcal{Z}]$, where $\mathcal{X}$, $\mathcal{Y}$ and $\mathcal{Z}$ are Banach spaces. Take T in $\mathcal{B}[\mathcal{X}, \mathcal{Y}]$, S in $\mathcal{B}[\mathcal{Y}, \mathcal{Z}]$, and prove the following propositions.

(a) $S_n \xrightarrow{u} S$ and $T_n \xrightarrow{u} T$ implies $S_n T_n \xrightarrow{u} ST$.

(b) $S_n \xrightarrow{s} S$ and $T_n \xrightarrow{s} T$ implies $S_n T_n \xrightarrow{s} ST$.

(c) $S_n \xrightarrow{w} S$ and $T_n \xrightarrow{s} T$ implies $S_n T_n \xrightarrow{w} ST$.

(d) $S_n \xrightarrow{u} S$ and $T_n \xrightarrow{w} T$ implies $S_n T_n \xrightarrow{w} ST$.

If $\mathcal{X}$ and $\mathcal{Y}$ are Hilbert spaces, then it follows by the Riesz Representation Theorem that $T_n \xrightarrow{w} T$ if and only if $\langle (T_n - T)x \, ; y \rangle \to 0$ for every x in $\mathcal{X}$ and every y in $\mathcal{Y}$; that is,

$$\langle T_n x \, , y \rangle \to \langle T x \, ; y \rangle \quad \text{for every} \quad x \in \mathcal{X} \quad \text{and every} \quad y \subset \mathcal{Y},$$

which is equivalent to saying that the scalar-valued sequence $\{\langle T_n x \, ; y \rangle\}$ converges in $\mathbb{F}$ for every x in $\mathcal{X}$ and every y in $\mathcal{Y}$. Now, if $\mathcal{X} = \mathcal{Y}$ is a *complex* Hilbert space, then $T_n \xrightarrow{w} T$ if and only if

$$\langle T_n x \, ; x \rangle \to \langle T x \, ; x \rangle \quad \text{for every} \quad x \in \mathcal{X}.$$

Again, this turns out to be equivalent to saying that the scalar sequence $\{\langle T_n x \, ; x \rangle\}$ converges in $\mathbb{C}$ for every x in $\mathcal{X}$. It is worth noticing that, if $\{T_n\}$ is a sequence of self-adjoint operators, then the above equivalent conditions also hold if $\mathcal{X} = \mathcal{Y}$ is a *real* Hilbert space (see e.g., [32, pp. 381–384,430]).

Problem 3.3. Let $\{T_n\}$ be a $\mathcal{B}[\mathcal{H}, \mathcal{K}]$-valued sequence, where $\mathcal{H}$ and $\mathcal{K}$ are Hilbert spaces, and take T in $\mathcal{B}[\mathcal{H}, \mathcal{K}]$. Show that

$$T_n \xrightarrow{u} T \quad \text{if and only if} \quad T_n^* \xrightarrow{u} T^*,$$

$$T_n \xrightarrow{w} T \quad \text{if and only if} \quad T_n^* \xrightarrow{w} T^*.$$

Remarks: Consider the setup of Problem 3.2. It is easy to show (even if $\mathcal{X} = \mathcal{Y} = \mathcal{Z}$ is a Hilbert space) that

$$S_n \xrightarrow{w} S \quad \text{and} \quad T_n \xrightarrow{u} T \quad \text{does not imply} \quad S_n T_n \xrightarrow{s} ST.$$

For instance, if $\{T_n\}$ is a nonzero constant sequence, say, $T_n = I$ for all n so that $T_n \xrightarrow{u} I$ trivially, and $\{S_n\}$ is a sequence that converges weakly to zero $(S_n \xrightarrow{w} O)$ but does not converge strongly, then $\{S_n T_n\}$ does not converge strongly. Of course, swapping S_n with T_n we get: $S_n \xrightarrow{u} S$ and $T_n \xrightarrow{w} T$ does not imply $S_n T_n \xrightarrow{s} ST$. It is also easy to show that

$$S_n \xrightarrow{s} S \quad \text{and} \quad T_n \xrightarrow{w} T \quad \text{does not imply} \quad S_n T_n \xrightarrow{w} ST.$$

Indeed, we shall see in Problem 5.1 an example of a Hilbert space isometry S_+ such that $S_+^{*n} \xrightarrow{s} O$, and so $S_+^n \xrightarrow{w} O$. However, $S_+^{*n} S_+^n \xrightarrow{w} O$ because $S_+^{*n} S_+^n = I$ for all n. Now consider the setup of Problem 3.3. The adjoint operation preserves uniform and weak convergences but does not preserve strong convergence. In fact, the same operator of Problem 5.1, being an isometry, is such that $S_+^n \xrightarrow{s} O$. Hence

$$T_n \xrightarrow{s} T \quad \text{does not imply} \quad T_n^* \xrightarrow{s} T^*.$$

Thus (as we shall see in Problem 5.1) there exist weakly convergent sequences that do not converge strongly. These concepts, however, coincide for sequences of nonnegative operators that converge to zero.

Problem 3.4. If $\{Q_n\}$ is a sequence of nonnegative operators on a Hilbert space, then

$$Q_n \xrightarrow{w} O \quad \text{if and only if} \quad Q_n \xrightarrow{s} O.$$

A sequence $\{T_n\}$ of operators in $\mathcal{B}[\mathcal{H}]$ is *increasing* if $T_n \leq T_{n+1}$ for every n, and *decreasing* if $T_{n+1} \leq T_n$ for every n. If it is increasing or decreasing, then we say that $\{T_n\}$ is *monotone*. The next two problems are rather useful consequences of Problem 3.4.

Problem 3.5. A bounded monotone sequence of self-adjoint operators converges strongly.

An operator T on a normed space $\mathcal{X}$ is *weakly stable* if the power sequence $\{T^n\}$ converges weakly to the null operator. That is, if $T^n \xrightarrow{w} O$. If $\mathcal{X}$ is a Hilbert space, this means $\langle T^n x \,; y \rangle \to 0$ for every $x, y \in \mathcal{X}$ or, equivalently, $\langle T^n x \,; x \rangle \to 0$ for every x in $\mathcal{X}$ if $\mathcal{X}$ is a *complex* Hilbert space. An operator $T \in \mathcal{B}[\mathcal{X}]$ is *power bounded* if $\sup_n \|T^n\| < \infty$. Clearly, strong stability implies weak stability and, if $\mathcal{X}$ is a Banach space, weak stability implies power boundedness.

Problem 3.6. Let $\mathcal{H}$ and $\mathcal{K}$ be Hilbert spaces and take any $T \in \mathcal{B}[\mathcal{H}, \mathcal{K}]$.

(a) Show that $T^{*n}T^n \xrightarrow{s} O \iff T^{*n}T^n \xrightarrow{w} O \iff T^n \xrightarrow{s} O.$

(b) Now conclude: The concepts of strong and weak stabilities coincide for a self-adjoint operator.

Let $\mathcal{X}$ and $\mathcal{Y}$ be normed spaces and let Θ be a subset of $\mathcal{B}[\mathcal{X}, \mathcal{Y}]$. According to the Closed Set Theorem (cf. Solution 1.1), Θ is closed if and only if every sequence of bounded linear transformations in Θ that converges in $\mathcal{B}[\mathcal{X}, \mathcal{Y}]$ has its limit in Θ. Equivalently, every Θ-valued uniformly convergent sequence has its (uniform) limit in Θ. In this case the set Θ is also called *uniformly closed* in $\mathcal{B}[\mathcal{X}, \mathcal{Y}]$. A subset Θ of $\mathcal{B}[\mathcal{X}, \mathcal{Y}]$ is *strongly closed* in $\mathcal{B}[\mathcal{X}, \mathcal{Y}]$ if every Θ-valued strongly convergent sequence has its (strong) limit in Θ, and *weakly closed* in $\mathcal{B}[\mathcal{X}, \mathcal{Y}]$ if every Θ-valued weakly convergent sequence has its (weak) limit in Θ. Since uniform convergence implies strong convergence, which in turn implies weak convergence, it follows that

Θ is weakly closed $\implies$ Θ is strongly closed $\implies$ Θ is uniformly closed.

Take any operator T on a normed space $\mathcal{X}$. Recall that the commutant of T, $\{T\}' = \{L \in \mathcal{B}[\mathcal{X}] : LT = TL\}$, is a unital subalgebra of the normed algebra $\mathcal{B}[\mathcal{H}]$. Indeed, $\{T\}'$ is a weakly closed unital subalgebra of $\mathcal{B}[\mathcal{H}]$.

Problem 3.7. Show that the commutant $\{T\}'$ is weakly closed in $\mathcal{B}[\mathcal{X}]$.

Let $\mathcal{B}^+[\mathcal{H}] = \{Q \in \mathcal{B}[\mathcal{H}] : Q \geq O\}$ be the collection of all nonnegative operators on a Hilbert space $\mathcal{H}$. If Q and R are nonnegative operators on $\mathcal{H}$, then it is readily verified that $\alpha Q + (1 - \alpha)R$ is again a nonnegative operator for every real $\alpha \in [0, 1]$ so that $\mathcal{B}^+[\mathcal{H}]$ is a *convex* subset of $\mathcal{B}[\mathcal{H}]$. Moreover, it is clear that αQ is nonnegative for every $\alpha \geq 0$ so that $\mathcal{B}^+[\mathcal{H}]$ is a *cone* in $\mathcal{B}[\mathcal{H}]$. Actually, $\mathcal{B}^+[\mathcal{H}]$ is a weakly closed convex cone in $\mathcal{B}[\mathcal{H}]$.

Problem 3.8. Show that $\mathcal{B}^+[\mathcal{H}]$ is weakly closed in $\mathcal{B}[\mathcal{H}]$.

Let $\mathcal{H}$ be a Hilbert space. If T is an operator in $\mathcal{B}[\mathcal{H}]$ and if there exists an operator S in $\mathcal{B}[\mathcal{H}]$ such that $S^2 = T$, then S is referred to as a *square root* of T. Nonnegative operators have a unique nonnegative square root. This is an important result for Hilbert space operators that will be applied throughout the text. A proof of it that does not need the Spectral Theorem is based on Problems 3.5 and 3.8 (see [22, pp. 64,254] or [32, p. 402]).

Square Root Theorem. *Every nonnegative operator Q in $\mathcal{B}[\mathcal{H}]$ has a unique nonnegative square root $Q^{\frac{1}{2}}$ in $\mathcal{B}[\mathcal{H}]$. Moreover, the square root $Q^{\frac{1}{2}}$ commutes with every operator in $\mathcal{B}[\mathcal{H}]$ that commutes with Q. Their norms are related by $\|Q^{\frac{1}{2}}\| = \|Q\|^{\frac{1}{2}}$, and their kernels and closure of their ranges coincide; $\mathcal{N}(Q^{\frac{1}{2}}) = \mathcal{N}(Q)$ and $\mathcal{R}(Q^{\frac{1}{2}})^- = \mathcal{R}(Q)^-$.*

Let $\mathcal{H}$ and $\mathcal{K}$ be Hilbert spaces. Take an arbitrary T in $\mathcal{B}[\mathcal{H}, \mathcal{K}]$, and recall that T^*T is a nonnegative operator in $\mathcal{B}[\mathcal{H}]$. Thus, according to the Square Root Theorem, set

$$|T| = (T^*T)^{\frac{1}{2}}$$

in $\mathcal{B}[\mathcal{H}]$, which is the unique nonnegative square root of $|T|^2 = T^*T$. The operator $|T|$ is referred to as the *absolute value* of T.

Problem 3.9. Take any T in $\mathcal{B}[\mathcal{H}, \mathcal{K}]$, where $\mathcal{H}$ and $\mathcal{K}$ are Hilbert spaces, and prove the following assertions.

(a) $\|T\| = \||T|^2\|^{\frac{1}{2}} = \||T|\| = \||T|^{\frac{1}{2}}\|^2$.

(b) $\langle |T|x \, ; x \rangle = \||T|^{\frac{1}{2}}x\|^2 \leq \||T|x\| \|x\|$ for every $x \in \mathcal{H}$.

(c) $\|Tx\|^2 = \||T|x\|^2 \leq \|T\| \langle |T|x \, ; x \rangle$ for every $x \in \mathcal{H}$.

Moreover, if $\mathcal{H} = \mathcal{K}$ (i.e., if $T \in \mathcal{B}[\mathcal{H}]$), then show that

(d) $\mathcal{B}^+[\mathcal{H}] = \{T \in \mathcal{B}[\mathcal{H}]: T = |T|\}$ (i.e., $T \geq O$ if and only if $T = |T|$),

(e) $T^n \xrightarrow{s} O \iff |T^n| \xrightarrow{s} O \iff |T^n| \xrightarrow{w} O$.

Solutions

Solution 3.1. Take an arbitrary bounded sequence $a = \{\alpha_k\}_{k=0}^{\infty}$ in ℓ_+^{∞} and let $D = \mathrm{diag}(\{\alpha_k\}_{k=0}^{\infty}): \ell_+^p \to \ell_+^p$ be a diagonal mapping.

(a) Since $a = \{\alpha_k\}_{k=0}^{\infty} \in \ell_+^{\infty}$,

$$\|Dx\| = \left(\sum_{k=0}^{\infty} |\alpha_k \xi_k|^p\right)^{\frac{1}{p}} \leq \sup_{k \geq 0} |\alpha_k| \left(\sum_{k=0}^{\infty} |\xi_k|^p\right)^{\frac{1}{p}} = \sup_{k \geq 0} |\alpha_k| \|x\|$$

for every $x = \{\xi_k\}_{k=0}^{\infty} \in \ell_+^p$, so that $\|D\| \leq \sup_k |\alpha_k| = \|a\|_{\infty}$. On the other hand, consider the ℓ_+^p-valued sequence $\{e_i\}_{i=0}^{\infty}$ where, for each $i \geq 0$, e_i is a scalar-valued sequence with just one nonzero entry (equal to 1) at the ith position (that is, $e_i = \{\delta_{ik}\}_{k=0}^{\infty}$ for every $i \geq 0$, with $\delta_{ik} = 1$ if $k = i$ and $\delta_{ik} = 0$ if $k \neq i$). Since $De_i = \alpha_i e_i$ and $\|e_i\| = 1$, $\|De_i\| = |\alpha_i|$ and hence $\|D\| = \sup_{\|x\|=1} \|Dx\| \geq \|De_i\| = |\alpha_i|$, for every $i \geq 0$. Thus $\|D\| \geq \sup_i |\alpha_i| = \|a\|_{\infty}$. Therefore,

$$\|D\| = \|a\|_{\infty}.$$

(b) It is readily verified by induction that the nth power of D is again a diagonal operator in $\mathcal{B}[\ell_+^p]$. Indeed,

$$D^n x = \{\alpha_k^n \xi_k\}_{k=0}^{\infty} \quad \text{for every} \quad x = \{\xi_k\}_{k=0}^{\infty} \in \ell_+^p$$

so that $\|D^n\| = \sup_k |\alpha_k|^n = \|a\|_\infty^n$ for every $n \geq 0$. Hence $D \in \mathcal{B}[\ell_+^p]$ is uniformly stable if and only if $\|a\|_\infty < 1$; that is,

$$D^n \xrightarrow{\ u\ } O \quad \text{if and only if} \quad \sup_k |\alpha_k| < 1.$$

(c) If $\|D^n x\| \to 0$ for every x in ℓ_+^p, then, in particular (see item (a) above), $\|D^n e_i\| = |\alpha_i|^n \to 0$ as $n \to \infty$, and hence $|\alpha_i| < 1$, for every $i \geq 0$. On the other hand, take an arbitrary vector $x = \{\xi_k\}_{k=0}^\infty$ in ℓ_+^p so that

$$0 \leq \|D^n x\|^p = \sum_{k=0}^\infty |\alpha_k|^{np} |\xi_k|^p = \sum_{k=0}^m |\alpha_k|^{np} |\xi_k|^p + \sum_{k=m+1}^\infty |\alpha_k|^{np} |\xi_k|^p$$

$$\leq \max_{0 \leq k \leq m} |\alpha_k|^{np} \sum_{k=0}^m |\xi_k|^p + \sup_{k>m} |\alpha_k|^{np} \sum_{k=m+1}^\infty |\xi_k|^p$$

for every pair of integers $m, n \geq 0$. If $|\alpha_k| < 1$ for every $k \geq 0$, then $\sup_k |\alpha_k|^{np} \leq 1$ for all $n \geq 0$. Thus, as $\sum_{k=0}^m |\xi_k|^p \leq \|x\|^p$ for all $m \geq 0$,

$$0 \leq \|D^n x\|^p \leq \max_{0 \leq k \leq m} |\alpha_k|^{np} \|x\|^p + \sum_{k=m+1}^\infty |\xi_k|^p$$

for every $m, n \geq 0$. Now take an arbitrary positive number, say $\varepsilon > 0$. Since $\sum_{k=m+1}^\infty |\xi_k|^p \to 0$ as $m \to \infty$ (for $\sum_{k=0}^\infty |\xi_k|^p = \|x\|^p < \infty$), it follows that there exists a positive integer m_ε such that

$$m \geq m_\varepsilon \quad \text{implies} \quad \sum_{k=m+1}^\infty |\xi_k|^p < \frac{\varepsilon^p}{2}.$$

Moreover, if $|\alpha_k| < 1$ for every $k \geq 0$, then $\max_{0 \leq k \leq m} |\alpha_k|^p < 1$ so that $\lim_n \max_{0 \leq k \leq m} |\alpha_k|^{np} = \lim_n (\max_{0 \leq k \leq m} |\alpha_k|^p)^n = 0$ for every nonnegative integer m. In particular, $\lim_n \max_{0 \leq k \leq m_\varepsilon} |\alpha_k|^{np} = 0$. Then there exists a positive integer n_ε such that

$$n \geq n_\varepsilon \quad \text{implies} \quad \max_{0 \leq k \leq m_\varepsilon} |\alpha_k|^{np} < \frac{\varepsilon^p}{2}.$$

But $0 \leq \|D^n x\|^p \leq \max_{0 \leq k \leq m_\varepsilon} |\alpha_k|^{np} \|x\|^p + \sum_{k=m_\varepsilon+1}^\infty |\xi_k|^p$, and so

$$n \geq n_\varepsilon \quad \text{implies} \quad 0 \leq \|D^n x\| < \varepsilon.$$

Therefore, if $|\alpha_k| < 1$ for every $k \geq 0$, then $\|D^n x\| \to 0$ for every x in ℓ_+^p. Hence $D \in \mathcal{B}[\ell_+^p]$ is strongly stable if and only if $|\alpha_k| < 1$ for every $k \geq 0$; that is,

$$D^n \xrightarrow{\ s\ } O \quad \text{if and only if} \quad |\alpha_k| < 1 \text{ for every } k \geq 0.$$

(d) If $\alpha_k = \frac{k+1}{k+2}$ for each $k \geq 0$, then the diagonal $D = \text{diag}(\{\alpha_k\}_{k=0}^{\infty})$ in $\mathcal{B}[\ell_+^p]$ is strongly stable (since $|\alpha_k| < 1$ for every $k \geq 0$) but D is not uniformly stable (because $\sup_k |\alpha_k| = 1$).

Solution 3.2. Let $\mathcal{X}$ and $\mathcal{Y}$ be Banach spaces, and let $\mathcal{Z}$ be a normed space. Suppose each S_n lies in $\mathcal{B}[\mathcal{Y}, \mathcal{Z}]$ and each T_n lies in $\mathcal{B}[\mathcal{X}, \mathcal{Y}]$. Take T in $\mathcal{B}[\mathcal{X}, \mathcal{Y}]$ and S in $\mathcal{B}[\mathcal{Y}, \mathcal{Z}]$. Recall that on a Banach space even weak convergence implies boundedness. To show that (a), (b) and (c) hold true, note that $S_n T_n - ST = S_n(T_n - T) + (S_n - S)T$. Indeed, take an arbitrary integer $n \geq 0$, an arbitrary x in $\mathcal{X}$, an arbitrary f in $\mathcal{Z}^*$, put $g = f \circ S$ in $\mathcal{Y}^*$ (i.e., $g(y) = f(Sy)$ for every $y \in \mathcal{Y}$), and observe that

(a) $\|S_n T_n - ST\| \leq \sup_n \|S_n\| \|T_n - T\| + \|S_n - S\| \|T\|$,

(b) $\|(S_n T_n - ST)x\| \leq \sup_n \|S_n\| \|(T_n - T)x\| + \|(S_n - S)Tx\|$,

(c) $|f((S_n T_n - ST)x)| \leq \|f\| \sup_n \|S_n\| \|(T_n - T)x\| + |f((S_n - S)Tx)|$.

Moreover, since $S_n T_n - ST = (S_n - S)T_n + S(T_n - T)$, (d) also holds true:

(d) $|f((S_n T_n - ST)x)| \leq \|f\| \|S_n - S\| \sup_n \|T_n\| \|x\| + |g((T_n - T)x)|$.

Solution 3.3. Let $\{T_n\}$ be a $\mathcal{B}[\mathcal{H}, \mathcal{K}]$-valued sequence, where $\mathcal{H}$ and $\mathcal{K}$ are Hilbert spaces. Since $\|T_n^* - T^*\| = \|(T_n - T)^*\| = \|T_n - T\|$ for every T in $\mathcal{B}[\mathcal{H}, \mathcal{K}]$, it follows that $T_n \xrightarrow{u} T$ if and only if $T_n^* \xrightarrow{u} T^*$. Similarly, since $\langle (T_n^* - T^*)x ; y \rangle = \langle (T_n - T)^* x ; y \rangle = \langle x ; (T_n - T)y \rangle = \overline{\langle (T_n - T)y ; x \rangle}$ for every $x \in \mathcal{H}$ and $y \in \mathcal{K}$, it follows that $T_n \xrightarrow{w} T$ if and only if $T_n^* \xrightarrow{w} T^*$.

Solution 3.4. Let $\{Q_n\}$ be a sequence of nonnegative operators acting on a Hilbert space $\mathcal{H}$. Recall that strong convergence always implies weak convergence (to the same limit). On the other hand, since $\mathcal{H}$ is a Hilbert space, $\sup_n \|Q_n\| < \infty$ if $\{Q_n\}$ converges weakly. Therefore, $Q_n \xrightarrow{w} O$ implies $Q_n \xrightarrow{s} O$ because

$$\|Q_n x\|^2 \leq \sup_k \|Q_k\| \langle Q_n x ; x \rangle$$

for every $x \in \mathcal{H}$ and every integer n, according to Problem 2.2.

Solution 3.5. Suppose $\{T_n\}$ is a bounded decreasing sequence of self-adjoint operators on a Hilbert space $\mathcal{H}$ and take an arbitrary x in $\mathcal{H}$. Note that each $\langle T_n x ; x \rangle$ lies in $\mathbb{R}$ because each T_n is self-adjoint. Moreover, since $\sup_n |\langle T_n x ; x \rangle| \leq \sup_n \|T_n\| \|x\|^2$ and since $\sup_n \|T_n\| < \infty$, it follows that $\{\langle T_n x ; x \rangle\}$ is bounded, and it is decreasing as $\{T_n\}$ is decreasing. Thus $\{\langle T_n x ; x \rangle\}$ is a real-valued bounded decreasing sequence, and therefore converges in $\mathbb{R}$, which implies that it converges in $\mathbb{F}$ (where $\mathbb{F}$ stands for the scalar field; either $\mathbb{F} = \mathbb{R}$ or $\mathbb{F} = \mathbb{C}$). But this means that $T_n - T \xrightarrow{w} O$ for some operator T on $\mathcal{H}$ (it does not matter whether $\mathcal{H}$ is a real or complex

Hilbert space because $\{T_n\}$ is a sequence of self-adjoint operators). Furthermore, as $\{T_n\}$ is decreasing, $\langle T_n x ; x \rangle \geq \langle T_{n+k} x ; x \rangle \to \langle Tx ; x \rangle$ as $k \to \infty$ so that $T_n - T \geq O$ for every n, and hence $T_n - T \xrightarrow{s} O$ according to Problem 3.4. Now suppose $\{T_n\}$ is a bounded increasing sequence of self-adjoint operators so that $\{-T_n\}$ is a bounded decreasing sequence of self-adjoint operators. The above argument ensures that $\{-T_n\}$ also converges strongly, and so does $\{T_n\}$. Summing up: if $\{T_n\}$ is a bounded monotone sequence of self-adjoint operators, then it converges strongly.

Solution 3.6. Take any $T \in \mathcal{B}[\mathcal{H}, \mathcal{K}]$, where $\mathcal{H}$ and $\mathcal{K}$ are Hilbert spaces.

(a) Since $T^{*n} T^n$ is a nonnegative operator on $\mathcal{H}$ for each $n \geq 1$, it follows from Problem 3.4 that

$$T^{*n} T^n \xrightarrow{s} O \quad \text{if and only if} \quad T^{*n} T^n \xrightarrow{w} O.$$

But $\|T^n x\|^2 = \langle T^{*n} T^n x ; x \rangle$ for every x in $\mathcal{H}$ and every $n \geq 1$. Thus

$$T^{*n} T^n \xrightarrow{w} O \quad \text{if and only if} \quad T^n \xrightarrow{s} O.$$

(b) It is clear that $T^n \xrightarrow{w} O$ implies $T^{2n} \xrightarrow{w} O$. Therefore, if $\mathcal{H} = \mathcal{K}$ and if T is a self-adjoint operator, then $T^n \xrightarrow{w} O$ implies $T^{*n} T^n \xrightarrow{w} O$, which in turn implies $T^n \xrightarrow{s} O$ according to item (a). Since strong stability always implies weak stability, we get:

$$\text{If } T^* = T, \text{ then } T^n \xrightarrow{w} O \iff T^n \xrightarrow{s} O.$$

Solution 3.7. Let $\mathcal{X}^*$ be the dual space of a Banach space $\mathcal{X}$. Take an arbitrary sequence $\{L_n\}$ of operators in $\{T\}'$ that converges weakly to L in $\mathcal{B}[\mathcal{X}]$. That is,

$$f\big((L_n - L)x\big) \to 0 \quad \text{for every} \quad x \in \mathcal{X} \quad \text{and every} \quad f \in \mathcal{X}^*.$$

Since each L_n lies in $\{T\}'$,

$$LT - TL = LT - L_n T + T L_n - TL = (L - L_n)T + T(L_n - L).$$

Take $f \in \mathcal{X}^*$ and $x \in \mathcal{X}$ arbitrary. Put $g = f \circ T$ in $\mathcal{X}^*$ and note that

$$f\big((LT - TL)x\big) = -f\big((L_n - L)Tx\big) + g\big((L_n - L)x\big) \to 0.$$

Thus $f((LT - TL)x) = 0$ for every x in $\mathcal{X}$ and every f in $\mathcal{X}^*$, which implies that $(LT - TL)x = 0$ for every x in $\mathcal{X}$ (by the Hahn–Banach Theorem; see e.g., [32, p. 265]). Therefore, $LT - TL = O$. Equivalently, $L \in \{T\}'$, and hence $\{T\}'$ is weakly closed in $\mathcal{B}[\mathcal{X}]$.

Solution 3.8. Take an arbitrary sequence $\{Q_n\}$ of nonnegative operators on a Hilbert space $\mathcal{H}$ that converges weakly to Q in $\mathcal{B}[\mathcal{H}]$. Since $Q_n \xrightarrow{w} Q$,

$$\langle Q_n x\,;x\rangle \to \langle Qx\,;x\rangle \quad \text{for every} \quad x \in \mathcal{H}.$$

But $\langle Qx\,;x\rangle$ is the limit of a sequence $\{\langle Q_n x\,;x\rangle\}$ of nonnegative real numbers, so that $\langle Qx\,;x\rangle$ is a nonnegative real number, for every x in $\mathcal{H}$. That is, $Q \geq O$. Thus $Q \in \mathcal{B}^+[\mathcal{H}]$, and hence $\mathcal{B}^+[\mathcal{H}]$ is weakly closed in $\mathcal{B}[\mathcal{H}]$.

Solution 3.9. Take any $T \in \mathcal{B}[\mathcal{H}, \mathcal{K}]$, where $\mathcal{H}$ and $\mathcal{K}$ are Hilbert spaces.

(a) $\|T\|^2 = \|T^*T\|$ and $\|Q^{\frac{1}{2}}\| = \|Q\|^{\frac{1}{2}}$ for every nonnegative Q in $\mathcal{B}[\mathcal{H}]$. Therefore, $\|T\| = \||T|^2\|^{\frac{1}{2}} = \||T|\| = \||T|^{\frac{1}{2}}\|^2$.

(b) $\||T|^{\frac{1}{2}}x\|^2 = \langle |T|^{\frac{1}{2}}x\,;|T|^{\frac{1}{2}}x\rangle = \langle |T|x\,;x\rangle \leq \||T|x\|\|x\|$ for every $x \in \mathcal{H}$ by the Schwarz inequality.

(c) $\|Tx\|^2 = \langle Tx\,;Tx\rangle = \langle T^*Tx\,;x\rangle = \langle |T|^2x\,;x\rangle = \langle |T|x\,;|T|x\rangle = \||T|x\|^2$ and $\||T|x\|^2 \leq \||T|^{\frac{1}{2}}\|^2\||T|^{\frac{1}{2}}x\|^2 = \|T\|\langle |T|x\,;x\rangle$ for every $x \in \mathcal{H}$.

(d) Clearly, $T = |T|$ implies $T \geq O$ because $|T|$ is always nonnegative. On the other hand, if T is self-adjoint, then $T^2 = T^*T = |T|^2 \geq O$ so that $(T^2)^{\frac{1}{2}} = |T|$ by uniqueness of the nonnegative square root of nonnegative operators. Using the same argument, and recalling that $T \geq O$ implies $T^2 \geq O$, we get $(T^2)^{\frac{1}{2}} = T$ whenever $T \geq O$. Thus, since nonnegative operators are self-adjoint, $T \geq O$ implies $T = |T|$.

(e) According to (c), $\||T^n|x\| = \|T^n x\|$ for every $x \in \mathcal{H}$ and every $n \geq 1$. Thus $T^n \xrightarrow{s} O$ implies $|T^n| \xrightarrow{s} O$, which trivially implies $|T^n| \xrightarrow{w} O$. Since $\|T^n x\|^2 \leq \||T^n|\|\langle |T^n|x\,;x\rangle$ for every $x \in \mathcal{H}$ and every $n \geq 1$ by (a) and (c), and since $\sup_n \||T^n|\| < \infty$ whenever $|T^n| \xrightarrow{w} O$ (because $\mathcal{H}$ is a Hilbert space), it follows that $|T^n| \xrightarrow{w} O$ implies $T^n \xrightarrow{s} O$.

4
Reducing Subspaces

If $\mathcal{M}$ is a linear manifold of an inner product space, then $\mathcal{M} \perp \mathcal{M}^\perp$, and therefore $\mathcal{M} \cap \mathcal{M}^\perp = \{0\}$. A central result of Hilbert space geometry (the Projection Theorem) says that, if $\mathcal{H}$ is a Hilbert space and $\mathcal{M}$ is a subspace of $\mathcal{H}$, then $\mathcal{M} + \mathcal{M}^\perp = \mathcal{H}$. In other words, *the orthogonal complement of a subspace $\mathcal{M}$ of a Hilbert space is a complementary subspace of $\mathcal{M}$* (see e.g., [32, pp. 339,368]). Thus, in a Hilbert space, every subspace has a complementary subspace, and this only happens in a Hilbert space [39].

If $\mathcal{M}$ and $\mathcal{N}$ are linear manifolds of a linear space $\mathcal{X}$, then the *direct sum* $\mathcal{M} \oplus \mathcal{N}$ is the linear space of all ordered pairs (u, v) with u in $\mathcal{M}$ and v in $\mathcal{N}$, where vector addition and scalar multiplication are defined coordinate-wise. The ordinary sum $\mathcal{M} + \mathcal{N}$ is the linear manifold of $\mathcal{X}$ consisting of all sums $u + v$ with u in $\mathcal{M}$ and v in $\mathcal{N}$. Although ordinary and direct sums are different linear spaces, they are isomorphic if $\mathcal{M} \cap \mathcal{N} = \{0\}$. Moreover, if $\mathcal{X}$ is an inner product space and if $\mathcal{M}$ and $\mathcal{N}$ are *orthogonal* complementary linear manifolds (i.e., if $\mathcal{X} = \mathcal{M} + \mathcal{N}$ and $\mathcal{M} \perp \mathcal{N}$), then $\mathcal{M} + \mathcal{N}$ and $\mathcal{M} \oplus \mathcal{N}$ are unitarily equivalent — ordinary and direct sums of *orthogonal* complementary linear manifolds are unitarily equivalent, where the inner product in $\mathcal{M} \oplus \mathcal{N}$ is given by $\langle (u_1, v_2)\,; (u_2, v_2) \rangle = \langle u_1\,; u_2 \rangle + \langle v_1\,; v_2 \rangle$. In this case, we may identify $\mathcal{M} + \mathcal{N}$ with $\mathcal{M} \oplus \mathcal{N}$ and write $\mathcal{X} = \mathcal{M} \oplus \mathcal{N}$.

Projection Theorem. *For every subspace $\mathcal{M}$ of any Hilbert space $\mathcal{H}$ there exists a unique orthogonal projection P in $\mathcal{B}[\mathcal{H}]$ such that $\mathcal{R}(P) = \mathcal{M}$ (and hence $\mathcal{N}(P) = \mathcal{M}^\perp$). Consequently, every Hilbert space $\mathcal{H}$ can be decomposed as an ordinary sum $\mathcal{H} = \mathcal{M} + \mathcal{M}^\perp$ or, equivalently, as an orthogonal direct sum $\mathcal{H} = \mathcal{M} \oplus \mathcal{M}^\perp$, for any subspace $\mathcal{M}$ of $\mathcal{H}$.*

It is worth noticing that, rigorously, $\mathcal{H} = \mathcal{M} + \mathcal{M}^\perp \cong \mathcal{M} \oplus \mathcal{M}^\perp$, where $\cong$ means *unitarily equivalent*. We shall, however, stick to the usual (and natural) identification and write $\mathcal{H} = \mathcal{M} \oplus \mathcal{M}^\perp$ instead of $\mathcal{H} \cong \mathcal{M} \oplus \mathcal{M}^\perp$. In this sense, vectors u in $\mathcal{M}$ and v in $\mathcal{M}^\perp$ will be identified with vectors $(u, 0)$ and $(0, v)$ in $\mathcal{M} \oplus \mathcal{M}^\perp$, respectively. Therefore, the sum $x = u + v$ in $\mathcal{M} + \mathcal{M}^\perp$ may be interpreted as $x = u \oplus v$ in $\mathcal{M} \oplus \mathcal{M}^\perp$ as a short expression for the direct sum $(u, 0) \oplus (0, v)$. Moreover, $\mathcal{M} \subseteq \mathcal{H}$ is identified with $\mathcal{M} \oplus \{0\}$ (which is a subspace of $\mathcal{M} \oplus \mathcal{M}^\perp$) and $\mathcal{M}^\perp \subseteq \mathcal{H}$ with $\{0\} \oplus \mathcal{M}^\perp$ (also a subspace of $\mathcal{M} \oplus \mathcal{M}^\perp$).

The unique orthogonal projection $P: \mathcal{H} \to \mathcal{H}$ for which $\mathcal{R}(P) = \mathcal{M}$ is called *the orthogonal projection onto* $\mathcal{M}$. Since $\mathcal{H} = \mathcal{M} \oplus \mathcal{M}^\perp$, it is usual to write $\mathcal{M}^\perp = \mathcal{H} \ominus \mathcal{M}$ for the orthogonal complement of $\mathcal{M}$.

Problem 4.1. Let $T \in \mathcal{B}[\mathcal{H}]$ be a nonzero operator acting on a Hilbert space $\mathcal{H}$. Show that the following assertions are pairwise equivalent.

(a) T has a nontrivial invariant subspace.

(b) There exists a nontrivial projection $P \in \mathcal{B}[\mathcal{H}]$ such that $PTP = TP$.

(c) The operator equation $STS = TS$ has a nontrivial solution (i.e., a solution $S \in \mathcal{B}[\mathcal{H}]$ such that $O \neq S \neq I$).

Let $\mathcal{M}$ be a subspace of a Hilbert space $\mathcal{H}$. An operator T on $\mathcal{H}$ can be written with respect to the decomposition $\mathcal{H} = \mathcal{M} \oplus \mathcal{M}^\perp$ as a 2×2 matrix with linear transformation entries,

$$T = \begin{pmatrix} A & B \\ C & D \end{pmatrix},$$

where $A \in \mathcal{B}[\mathcal{M}]$, $B \in \mathcal{B}[\mathcal{M}^\perp, \mathcal{M}]$, $C \in \mathcal{B}[\mathcal{M}, \mathcal{M}^\perp]$ and $D \in \mathcal{B}[\mathcal{M}^\perp]$. Note that, with respect to the same decomposition, the orthogonal projection onto $\mathcal{M}$ is written as $P = \begin{pmatrix} I & O_1 \\ O_2 & O \end{pmatrix}$, where O is the null operator on $\mathcal{M}^\perp$ and I is the identity operator on $\mathcal{M}$. Moreover, O_1 and O_2 also stand for null transformations, now of $\mathcal{M}^\perp$ into $\mathcal{M}$ and of $\mathcal{M}$ into $\mathcal{M}^\perp$, respectively. We shall, of course, drop the subscripts and write

$$P = \begin{pmatrix} I & O \\ O & O \end{pmatrix}.$$

If T is an operator in $\mathcal{B}[\mathcal{H}]$ and $\mathcal{M}$ is T-invariant (i.e., $T(\mathcal{M}) \subseteq \mathcal{M}$), then the restriction of T to $\mathcal{M}$, $T|_{\mathcal{M}}$ in $\mathcal{B}[\mathcal{M}, \mathcal{H}]$, is such that $\mathcal{R}(T|_{\mathcal{M}}) \subseteq \mathcal{M}$. Thus $T|_{\mathcal{M}}: \mathcal{M} \to \mathcal{M}$ and we shall consider $T|_{\mathcal{M}}$ as an operator in $\mathcal{B}[\mathcal{M}]$.

Problem 4.2. Let T be an operator on a Hilbert space $\mathcal{H}$ and let $\mathcal{M}$ be a subspace of $\mathcal{H}$. Show that, if $\mathcal{M}$ is T-invariant, then T can be written with respect to the decomposition $\mathcal{H} = \mathcal{M} \oplus \mathcal{M}^\perp$ as

$$T = \begin{pmatrix} T|_{\mathcal{M}} & B \\ O & D \end{pmatrix},$$

where B lies in $\mathcal{B}[\mathcal{M}^\perp, \mathcal{M}]$, D lies in $\mathcal{B}[\mathcal{M}^\perp]$, and $T|_\mathcal{M}$ is the restriction of T to $\mathcal{M}$, which lies in $\mathcal{B}[\mathcal{M}]$. Conversely, take $A \in \mathcal{B}[\mathcal{M}]$, $B \in \mathcal{B}[\mathcal{M}^\perp, \mathcal{M}]$ and $D \in \mathcal{B}[\mathcal{M}^\perp]$. If

$$T = \begin{pmatrix} A & B \\ O & D \end{pmatrix}$$

on $\mathcal{H} = \mathcal{M} \oplus \mathcal{M}^\perp$, then $\mathcal{M}$ is T-invariant and $A = T|_\mathcal{M}$.

Problem 4.3. Let T be an operator on a Hilbert space $\mathcal{H}$ and let $\mathcal{M}$ be a subspace of $\mathcal{H}$. Show that

(a) $\mathcal{M}$ is T-invariant if and only if $\mathcal{M}^\perp$ is T^*-invariant,

(b) $\mathcal{M}$ is invariant for every operator that commutes with T if and only if $\mathcal{M}^\perp$ is invariant for every operator that commutes with T^*.

Let $\mathcal{M}$ be a subspace of a Hilbert space $\mathcal{H}$. Observe that $\mathcal{M}$ is nontrivial if and only if $\mathcal{M}^\perp$ is. That is, $\{0\} \neq \mathcal{M} \neq \mathcal{H}$ if and only if $\{0\} \neq \mathcal{M}^\perp \neq \mathcal{H}$. In fact, since $\mathcal{H} = \mathcal{M} + \mathcal{M}^\perp$, we get $\mathcal{M} = \{0\}$ if and only if $\mathcal{M}^\perp = \mathcal{H}$, and $\mathcal{M}^\perp = \{0\}$ if and only if $\mathcal{M} = \mathcal{H}$. Therefore, according to Problem 4.3, *an operator on a Hilbert space has a nontrivial invariant (hyperinvariant) subspace if and only if its adjoint has.*

Let T be an operator on Hilbert space $\mathcal{H}$ and let $\mathcal{M}$ be a subspace of $\mathcal{H}$. If $\mathcal{M}$ and its orthogonal complement $\mathcal{M}^\perp$ are both T-invariant (i.e., if $T(\mathcal{M}) \subseteq \mathcal{M}$ and $T(\mathcal{M}^\perp) \subseteq \mathcal{M}^\perp$), then we say that $\mathcal{M}$ *reduces* T (or $\mathcal{M}$ is a *reducing subspace* for T). If $\mathcal{M}$ reduces T and $\{0\} \neq \mathcal{M} \neq \mathcal{H}$, then $\mathcal{M}$ is a *nontrivial reducing subspace* for T. It is clear that if T has a nontrivial reducing (or just invariant) subspace, then the dimension of $\mathcal{H}$ is greater than 1. An operator T acting on a Hilbert space $\mathcal{H}$ of dimension greater than 1 is *reducible* if it has a nontrivial reducing subspace. Recall that $\mathcal{M} \neq \{0\}$ if and only if $\mathcal{M}^\perp \neq \mathcal{H}$, and $\mathcal{M}^\perp \neq \{0\}$ if and only if $\mathcal{M} \neq \mathcal{H}$. Thus an operator T on $\mathcal{H}$ is reducible if there exists a subspace $\mathcal{M}$ of $\mathcal{H}$ such that both $\mathcal{M}$ and $\mathcal{M}^\perp$ are nonzero and T-invariant or, equivalently, if $\mathcal{M}$ is nontrivial and invariant for both T and T^*, as we shall see next.

Problem 4.4. A subspace $\mathcal{M}$ of a Hilbert space $\mathcal{H}$ reduces $T \in \mathcal{B}[\mathcal{H}]$ if and only if $\mathcal{M}$ is invariant for both T and T^*.

Problem 4.5. Let L and T be operators on a Hilbert space. Show that, if L commutes with both T and T^*, then $\mathcal{N}(L)$ and $\mathcal{R}(L)^-$ reduce T.

Problem 4.6. Let T be an operator on a Hilbert space. Show that

(a) if T commutes with an orthogonal projection P, then $\mathcal{R}(P)$ is a reducing subspace for T;

(b) T is reducible if and only if it commutes with a nontrivial orthogonal projection.

Problem 4.7. Let $\mathcal{M}$ be a subspace of a Hilbert space $\mathcal{H}$. Take any T in $\mathcal{B}[\mathcal{H}]$ and consider the orthogonal projection $P \colon \mathcal{H} \to \mathcal{H}$ onto $\mathcal{M}$. Prove the following assertions on the restrictions of T and T^* to $\mathcal{M}$.

(a) If $\mathcal{M}$ is T-invariant, then $(T|_{\mathcal{M}})^* = PT^*|_{\mathcal{M}}$.

(b) If $\mathcal{M}$ reduces T, then $(T|_{\mathcal{M}})^* = T^*|_{\mathcal{M}}$.

Problem 4.8. Let T be an operator on a Hilbert space $\mathcal{H}$ and let $\mathcal{M}$ be a subspace of $\mathcal{H}$. Show that, if $\mathcal{M}$ reduces T, then T can be written with respect to the decomposition $\mathcal{H} = \mathcal{M} \oplus \mathcal{M}^\perp$ as

$$ T = \begin{pmatrix} T|_{\mathcal{M}} & O \\ O & T|_{\mathcal{M}^\perp} \end{pmatrix}, $$

where $T|_{\mathcal{M}} \in \mathcal{B}[\mathcal{M}]$ is the restriction of T to $\mathcal{M}$ and $T|_{\mathcal{M}^\perp} \in \mathcal{B}[\mathcal{M}^\perp]$ is the restriction of T to $\mathcal{M}^\perp$. Conversely, take $A \in \mathcal{B}[\mathcal{M}]$ and $D \in \mathcal{B}[\mathcal{M}^\perp]$. If

$$ T = \begin{pmatrix} A & O \\ O & D \end{pmatrix} $$

on $\mathcal{H} = \mathcal{M} \oplus \mathcal{M}^\perp$, then $\mathcal{M}$ reduces T, $A = T|_{\mathcal{M}}$ and $D = T|_{\mathcal{M}^\perp}$.

If $T = \left(\begin{smallmatrix} A & O \\ O & D \end{smallmatrix}\right)$ on $\mathcal{H} = \mathcal{M} \oplus \mathcal{M}^\perp$, then we write $T = A \oplus D$ and say that T is the (orthogonal) *direct sum* of A and D. The operators $A \in \mathcal{B}[\mathcal{M}]$ and $D \in \mathcal{B}[\mathcal{M}^\perp]$ are referred to as *direct summands* of T. Thus what Problem 4.8 says is: *An operator T on a Hilbert space $\mathcal{H}$ is reducible if and only if it is the* (orthogonal) *direct sum of two operators on nonzero subspaces of $\mathcal{H}$, namely, $T = T|_{\mathcal{M}} \oplus T|_{\mathcal{M}^\perp}$ where both $\mathcal{M}$ and $\mathcal{M}^\perp$ are nonzero.* If an operator is not reducible, then it is called *irreducible*.

If $T = \left(\begin{smallmatrix} A & B \\ C & D \end{smallmatrix}\right)$ on $\mathcal{H} = \mathcal{M} \oplus \mathcal{M}^\perp$, then $T^* = \left(\begin{smallmatrix} A^* & C^* \\ B^* & D^* \end{smallmatrix}\right)$. In particular, if $T = A \oplus D$, then $T^* = A^* \oplus D^*$ and, as we already knew from Problem 4.4, T is reducible if and only if T^* is.

Problem 4.9. Every operator unitarily equivalent to a reducible operator is reducible. That is, show that reducibility is preserved under unitary equivalence.

Problem 4.10. Let T be an operator on a Hilbert space $\mathcal{H}$ and let $\mathcal{M}$ be a nonzero subspace of $\mathcal{H}$. Suppose $\|T\| \leq 1$. Show that, if $\mathcal{M}$ is T-invariant and $T|_{\mathcal{M}}$ is unitary, then $\mathcal{M}$ reduces T.

We close this chapter with the notion of infinite orthogonal direct sum. Let $\{\mathcal{X}_k\}$ be a nonempty collection of linear manifolds of a linear space $\mathcal{X}$. The *sum* $\sum_k \mathcal{X}_k$ is the linear manifold of $\mathcal{X}$ consisting of all *finite* sums of vectors in $\mathcal{X}$ with each summand being a vector in one of the linear manifolds $\mathcal{X}_k$. Moreover,

$$\sum_k \mathcal{X}_k = \operatorname{span}_k \mathcal{X}_k,$$

where $\operatorname{span}_k \mathcal{X}_k$ denotes the linear manifold of $\mathcal{X}$ spanned by $\{\mathcal{X}_k\}$; that is, $\operatorname{span}_k \mathcal{X}_k = \operatorname{span}\left(\bigcup_k \mathcal{X}_k\right)$ is the linear manifold of $\mathcal{X}$ made up of all (finite) linear combinations of vectors in $\bigcup_k \mathcal{X}_k$. Now let $\{\mathcal{X}_k\}$ be a collection of *subspaces* of a normed space $\mathcal{X}$ and set $\bigvee_k \mathcal{X}_k = \left(\operatorname{span}_k \mathcal{X}_k\right)^-$, the closure of $\operatorname{span}_k \mathcal{X}_k$, which coincides with the closure of $\sum_k \mathcal{X}_k$; that is,

$$\bigvee_k \mathcal{X}_k = \left(\sum_k \mathcal{X}_k\right)^-.$$

This is called the *topological sum* of $\{\mathcal{X}_k\}$. Recall that, even if $\{\mathcal{X}_k\}$ is a finite collection and $\mathcal{X}$ is a Banach space (so that each $\mathcal{X}_k$ is a Banach space), the topological sum does not necessarily coincide with the (ordinary) sum — the sum of a pair of closed linear manifolds is not necessarily closed. However, if $\{\mathcal{H}_k\}$ is a collection of *orthogonal* subspaces of a Hilbert space $\mathcal{H}$ (so that each $\mathcal{H}_k$ is itself a Hilbert space and $\mathcal{H}_j \perp \mathcal{H}_k$ whenever $j \neq k$), then every *finite* sum of subspaces is closed, and hence every *finite* sum of *orthogonal* subspaces coincides with its *finite* topological sum (this fails for infinite sums of orthogonal subspaces). Let $\bigoplus_k \mathcal{H}_k$ denote the (orthogonal) *direct sum* of $\{\mathcal{H}_k\}$, which is the collection of all square-summable families of vectors in $\mathcal{H}$ with each vector in each $\mathcal{H}_k$. This is a Hilbert space whose inner product is given by the sum of the inner products in each $\mathcal{H}_k$. If $\{\mathcal{H}_k\}$ is a countably infinite family, its orthogonal direct sum is the collection of all square-summable sequences of vectors in $\mathcal{H}$ with each vector in each $\mathcal{H}_k$ (recall that, if a sequence of pairwise orthogonal vectors in a Hilbert space is square-summable, then it yields an unconditional convergent series). The direct sum of pairwise orthogonal subspaces of a Hilbert space is unitarily equivalent to its topological sum;

$$\bigoplus_k \mathcal{H}_k \cong \bigvee_k \mathcal{H}_k = \left(\sum_k \mathcal{H}_k\right)^-.$$

According to the above unitary equivalence, it is usual to identify the direct sum with the topological sum, and write $\bigoplus_k \mathcal{H}_k = \bigvee_k \mathcal{H}_k$. If the orthogonal collection $\{\mathcal{H}_k\}$ *spans* $\mathcal{H}$ (i.e., if $\mathcal{H} = \bigvee_k \mathcal{H}_k$), then we write $\mathcal{H} = \bigoplus_k \mathcal{H}_k$. (See e.g., [32, §2.2,4.3,5.5–5.7].) Finally, let $\{T_k\}$ be a (similarly indexed) family of operators on $\mathcal{H}_k$. Suppose each $\mathcal{H}_k$ is nonzero and $\{T_k\}$ is bounded (i.e., $\sup_k \|T_k\| < \infty$), and consider the mapping $T: \bigoplus_k \mathcal{H}_k \to \bigoplus_k \mathcal{H}_k$ such that $T\{x_k\} = \{T_k x_k\}$ for every $\{x_k\}$ in $\bigoplus_k \mathcal{H}_k$. Actually, this defines an operator on the Hilbert space $\bigoplus_k \mathcal{H}_k$, which is denoted by

$$T = \bigoplus_k T_k.$$

and referred to as the *direct sum* of $\{T_k\}$. In fact, $\|T\| = \sup_k \|T_k\|$. Each $\mathcal{H}_k$ is unitarily equivalent to the subspace of $\bigoplus_k \mathcal{H}_k$ containing $\mathcal{H}_k$ as the only nonzero direct summand $(\mathcal{H}_k \cong \bigoplus_{j<k}\{0\} \oplus \mathcal{H}_k \oplus \bigoplus_{j>k}\{0\})$ at the kth position. Thus we may interpret $\mathcal{H}_k$ as a subspace of $\bigoplus_k \mathcal{H}_k$ and, in this case, each $\mathcal{H}_k$ reduces T. Indeed, $T|_{\mathcal{H}_k} = T_k$ in $\mathcal{B}[\mathcal{H}_k]$ is referred to as a *direct summand* of T in $\mathcal{B}[\bigoplus_k \mathcal{H}_k]$ so that $T = \bigoplus_k T|_{\mathcal{H}_k}$. Moreover, its adjoint is given by $T^* = \bigoplus_k T_k^* = \bigoplus_k T^*|_{\mathcal{H}_k}$ (see e.g., [32, p. 421]).

Solutions

Solution 4.1. Let $\mathcal{M}$ be a subspace of a Hilbert space $\mathcal{H}$, take any T in $\mathcal{B}[\mathcal{H}]$, consider the orthogonal projection $P \in \mathcal{B}[\mathcal{H}]$ onto $\mathcal{M}$, and take an arbitrary x in $\mathcal{X}$. Since $\mathcal{H} = \mathcal{M} + \mathcal{M}^\perp$, there exist u in $\mathcal{M}$ and v in $\mathcal{M}^\perp$ such that $x = u + v$ (u and v are unique because $\mathcal{M} \cap \mathcal{M}^\perp = \{0\}$). Thus,

$$PTPx = PTP(u + v) = PTPu + PTPv = PTu$$

(reason: $\mathcal{M}^\perp = \mathcal{N}(P)$ and $\mathcal{M} = \mathcal{R}(P)$, which is the set of all fixed points of P). If $\mathcal{M}$ is T-invariant, then and Tu lies in $\mathcal{M} = \mathcal{R}(P)$ so that

$$PTu = Tu = TPu = TP(x - v) = TPx - TPv = TPx.$$

Therefore, $PTP = TP$. Conversely, if this identity holds, then $\mathcal{M} = \mathcal{R}(P)$ is clearly T-invariant. Since $\{0\} \neq \mathcal{R}(P) \neq \mathcal{H}$ if and only if $O \neq P \neq I$, it follows that $\mathcal{M} = \mathcal{R}(P)$ is nontrivial if and only if P is. Hence (a) is equivalent to (b); that is, *T has a nontrivial invariant subspace $\mathcal{M}$ if and only if the orthogonal projection $P \in \mathcal{B}[\mathcal{H}]$ onto $\mathcal{M}$ is such that $PTP = TP$.* Finally, (b) implies (c) trivially and (c) implies (a) by Problem 1.4.

Solution 4.2. Let $\mathcal{M}$ be a subspace of a Hilbert space $\mathcal{H}$ and take any T on $\mathcal{H}$. Write T with respect to the decomposition $\mathcal{H} = \mathcal{M} \oplus \mathcal{M}^\perp$,

$$T = \begin{pmatrix} A & B \\ C & D \end{pmatrix},$$

with $A \in \mathcal{B}[\mathcal{M}]$, $B \in \mathcal{B}[\mathcal{M}^\perp, \mathcal{M}]$, $C \in \mathcal{B}[\mathcal{M}, \mathcal{M}^\perp]$ and $D \in \mathcal{B}[\mathcal{M}^\perp]$. If $\mathcal{M}$ is T-invariant, then $C: \mathcal{M} \to \mathcal{M}^\perp$ must be null because $\mathcal{M} \cap \mathcal{M}^\perp = \{0\}$. Now suppose $C = O$ and identify an arbitrary vector u in $\mathcal{M}$ with $(u, 0)$ in $\mathcal{M} \oplus \mathcal{M}^\perp$ so that $Tu = \begin{pmatrix} A & B \\ O & D \end{pmatrix}(u, 0) = (Au, 0) = Au \in \mathcal{M}$. Hence $\mathcal{M}$ is T-invariant and $T|_{\mathcal{M}} = A$.

Solution 4.3. Let $\mathcal{M}$ be a subspace of a Hilbert space $\mathcal{H}$ and let T be any operator in $\mathcal{B}[\mathcal{H}]$.

(a) Take an arbitrary y in $\mathcal{M}^\perp$. If $Tx \in \mathcal{M}$ whenever $x \in \mathcal{M}$, then $\langle x\,;T^*y\rangle = \langle Tx\,;y\rangle = 0$ for every $x \in \mathcal{M}$, and therefore $T^*y \perp \mathcal{M}$, which means that T^*y lies in $\mathcal{M}^\perp$. Conclusion: $T(\mathcal{M}) \subseteq \mathcal{M}$ implies $T^*(\mathcal{M}^\perp) \subseteq \mathcal{M}^\perp$. Conversely, since this holds for every operator in $\mathcal{B}[\mathcal{H}]$, it follows that $T^*(\mathcal{M}^\perp) \subseteq \mathcal{M}^\perp$ implies $T^{**}(\mathcal{M}^{\perp\perp}) \subseteq \mathcal{M}^{\perp\perp}$. But $T^{**} = T$ and $\mathcal{M}^{\perp\perp} = \mathcal{M}^- = \mathcal{M}$, and hence $T^*(\mathcal{M}^\perp) \subseteq \mathcal{M}^\perp$ implies $T(\mathcal{M}) \subseteq \mathcal{M}$. Summing up: $\mathcal{M}$ is T-invariant if and only if $\mathcal{M}^\perp$ is T^*-invariant.

(b) Let $\{T\}'$ be the commutant of T; the set of all operators in $\mathcal{B}[\mathcal{H}]$ that commute with T. It is clear that $L \in \{T\}'$ if and only if $L^* \in \{T^*\}'$. Suppose $\mathcal{M}$ is invariant for every operator that commutes with T, which means that $\mathcal{M}$ is L-invariant whenever $L \in \{T\}'$. This implies that $\mathcal{M}^\perp$ is L^*-invariant whenever $L \in \{T\}'$, according to (a). That is, $\mathcal{M}^\perp$ is L^*-invariant whenever $L^* \in \{T^*\}'$. Thus $\mathcal{M}^\perp$ is invariant for every operator that commutes with T^*. Dually, if $\mathcal{M}^\perp$ is invariant for every operator that commutes with T^*, then $\mathcal{M}^{\perp\perp} = \mathcal{M}^- = \mathcal{M}$ is invariant for every operator that commutes with $T^{**} = T$.

Solution 4.4. Since $\mathcal{M}^{\perp\perp} = \mathcal{M}^- = \mathcal{M}$, it follows by Problem 4.3(a) that $T(\mathcal{M}^\perp) \subseteq \mathcal{M}^\perp$ if and only if $T^*(\mathcal{M}) \subseteq \mathcal{M}$. Thus $T(\mathcal{M}) \subseteq \mathcal{M}$ and $T(\mathcal{M}^\perp) \subseteq \mathcal{M}^\perp$ if and only if $T(\mathcal{M}) \subseteq \mathcal{M}$ and $T^*(\mathcal{M}) \subseteq \mathcal{M}$. That is, $\mathcal{M}$ reduces T if and only if $\mathcal{M}$ is invariant for both T and T^*.

Solution 4.5. Let $\mathcal{H}$ be a Hilbert space and take T and L in $\mathcal{B}[\mathcal{H}]$. Problem 1.2 says that, if L commutes with T, then $\mathcal{N}(L)$ and $\mathcal{R}(L)^-$ are T-invariant. Similarly, if L commutes with T^*, then $\mathcal{N}(L)$ and $\mathcal{R}(L)^-$ are T^*-invariant. Therefore, according to Problem 4.4, if L commutes with T and with T^*, then both subspaces of $\mathcal{H}$, $\mathcal{N}(L)$ and $\mathcal{R}(L)^-$, reduce T.

Solution 4.6. Let $\mathcal{H}$ be a Hilbert space. Take an operator $T \in \mathcal{B}[\mathcal{H}]$ and an orthogonal projection $P \in \mathcal{B}[\mathcal{H}]$. Recall that $\mathcal{R}(P)$ is a subspace.

(a) If $PT = TP$, then it is clear that $\mathcal{R}(P)$ is T-invariant. Moreover, since P is self-adjoint, it follows that $PT^* = T^*P$, and hence $\mathcal{R}(P)$ is T^*-invariant. Therefore, $\mathcal{R}(P)$ reduces T (Problem 4.4).

(b) Recall: $\{0\} \neq \mathcal{R}(P) \neq \mathcal{H}$ if and only if $O \neq P \neq I$. Thus, according to item (a), if T commutes with a nontrivial orthogonal projection, then $\mathcal{R}(P)$ is a nontrivial reducing subspace for T; that is, T is reducible. Conversely, suppose T is reducible so that there exists a nontrivial subspace $\mathcal{M}$ such that both $\mathcal{M}$ and $\mathcal{M}^\perp$ are T-invariant. Since $\mathcal{M}$ is T-invariant, it follows that the nontrivial orthogonal projection P onto $\mathcal{M}$ is such that $PTP = TP$ (see Solution 4.1). Similarly, since $\mathcal{M}^\perp$ is T-invariant, it also follows that the complementary projection $E = I - P$ onto $\mathcal{M}^\perp$ is such that $ETE = TE$, and hence $PTP = PT$. Therefore, $PT = TP$.

Solution 4.7. Let $\mathcal{M}$ be an invariant subspace for T so that $T(\mathcal{M}) \subseteq \mathcal{M}$, and let P be the orthogonal projection onto $\mathcal{M}$. Since $Pv = v$ for every v in $\mathcal{M}$, and since P is self-adjoint (Problem 2.7), we get

$$\langle (T|_\mathcal{M})^* u ; v \rangle = \langle u ; T|_\mathcal{M} v \rangle = \langle u ; Tv \rangle$$
$$= \langle u ; TPv \rangle = \langle PT^* u ; v \rangle = \langle PT^*|_\mathcal{M} u ; v \rangle$$

for every $u, v \in \mathcal{M}$, and hence $(T|_\mathcal{M})^* = PT^*|_\mathcal{M}$. This proves (a). If $\mathcal{M}$ reduces T, then $\mathcal{M}$ is invariant for both T and T^* (Problem 4.4). Hence $\mathcal{R}(T^*|_\mathcal{M}) \subseteq \mathcal{M}$ so that $PT^*|_\mathcal{M} = T^*|_\mathcal{M}$ because $\mathcal{R}(P) = \mathcal{M}$. Therefore, $(T|_\mathcal{M})^* = T^*|_\mathcal{M}$, by item (a), which proves item (b).

Solution 4.8. Let $\mathcal{M}$ be a subspace of a Hilbert space $\mathcal{H}$ and take any T on $\mathcal{H}$. Write T with respect to the decomposition $\mathcal{H} = \mathcal{M} \oplus \mathcal{M}^\perp$,

$$T = \begin{pmatrix} A & B \\ C & D \end{pmatrix},$$

with $A \in \mathcal{B}[\mathcal{M}]$, $B \in \mathcal{B}[\mathcal{M}^\perp, \mathcal{M}]$, $C \in \mathcal{B}[\mathcal{M}, \mathcal{M}^\perp]$ and $D \in \mathcal{B}[\mathcal{M}^\perp]$. If both $\mathcal{M}$ and $\mathcal{M}^\perp$ are T-invariant, then $B \colon \mathcal{M}^\perp \to \mathcal{M}$ and $C \colon \mathcal{M} \to \mathcal{M}^\perp$ are both null because $\mathcal{M} \cap \mathcal{M}^\perp = \{0\}$. Now suppose $B = O$, $C = O$ and take arbitrary vectors u in $\mathcal{M}$ and v in $\mathcal{M}$. Identify them with $(u, 0)$ and $(0, v)$ in $\mathcal{M} \oplus \mathcal{M}^\perp$, respectively, and note that $\mathcal{M}$ and $\mathcal{M}^\perp$ are both T-invariant. Indeed, $Tu = \begin{pmatrix} A & O \\ O & D \end{pmatrix}(u, 0) = Au \in \mathcal{M}$ and $Tv = \begin{pmatrix} A & O \\ O & D \end{pmatrix}(0, v) = Dv \in \mathcal{M}^\perp$, so that $T|_\mathcal{M} = A$ and $T|_{\mathcal{M}^\perp} = D$.

Solution 4.9. Let $\mathcal{H}$ and $\mathcal{K}$ be unitarily equivalent Hilbert spaces. Take T and P in $\mathcal{B}[\mathcal{H}]$ and an arbitrary unitary transformation $U \colon \mathcal{K} \to \mathcal{H}$. Put $S = U^* TU$ and $E = U^* PU$ in $\mathcal{B}[\mathcal{K}]$. The operator E is an orthogonal projection if and only if P is. Indeed, $E^2 = U^* P^2 U$ and $E^* = U^* P^* U$ so that (see Problem 2.7) $E = E^2$ if and only if $P = P^2$ and $E = E^*$ if and only if $P = P^*$ (this equivalence might not hold if S and T were just similar). Moreover, $E = U^* PU$ is nontrivial if and only if P is, and E commutes with S if and only if P commutes with T (reason: $ES - SE = U^*(PT - TP)U$). Therefore, according to Problem 4.6, S is reducible if and only if T is.

Solution 4.10. $\|T\| \le 1$ if and only if $\|T^* T\| \le 1$ (reason: $\|T^* T\| = \|T\|^2$). If $\mathcal{M} \ne \{0\}$ is an invariant subspace for T and $U = T|_\mathcal{M}$ is unitary, then

$$T = \begin{pmatrix} U & B \\ O & D \end{pmatrix} \quad \text{so that} \quad T^* T = \begin{pmatrix} I & U^* B \\ B^* U & B^* B + D^* D \end{pmatrix}$$

with respect to the decomposition $\mathcal{H} = \mathcal{M} \oplus \mathcal{M}^\perp$ (cf. Problems 2.6 and 4.2). If $\|T\| \le 1$, then $\|u\|^2 + \|B^* U u\|^2 = \|T^* T(u, 0)\|^2 \le \|(u, 0)\|^2 = \|u\|^2$ for every u in $\mathcal{M}$. This implies that $B^* U = O$, and therefore $B^* = O$ (U is invertible) so that $B = O$. Thus $T = U \oplus D$, and hence $\mathcal{M}$ reduces T.

5
Shifts

An operator S_+ on a Hilbert space $\mathcal{H}$ is a *unilateral shift* if there exists an infinite sequence $\{\mathcal{H}_k\}_{k=0}^{\infty}$ of nonzero pairwise orthogonal subspaces of $\mathcal{H}$ such that $\mathcal{H} = \bigoplus_{k=0}^{\infty} \mathcal{H}_k$ (i.e., the orthogonal family $\{\mathcal{H}_k\}_{k=0}^{\infty}$ spans $\mathcal{H}$) and S_+ maps each $\mathcal{H}_k$ isometrically onto $\mathcal{H}_{k+1}$. Two Hilbert spaces are unitarily equivalent if and only if they have the same dimension (see e.g., [32, p. 365]). Since $S_+|_{\mathcal{H}_k} : \mathcal{H}_k \to \mathcal{H}_{k+1}$ is unitary (a surjective isometry), it follows that $\dim \mathcal{H}_{k+1} = \dim \mathcal{H}_k$, for every $k \geq 0$. This constant dimension is the *multiplicity* of S_+. The adjoint $S_+^* \in \mathcal{B}[\mathcal{H}]$ of $S_+ \in \mathcal{B}[\mathcal{H}]$ is referred to as a *backward unilateral shift*, also denoted by S_-. Writing $\bigoplus_{k=0}^{\infty} x_k$ for $\{x_k\}_{k=0}^{\infty}$ in $\bigoplus_{k=0}^{\infty} \mathcal{H}_k$, it follows that S_+ and S_+^* are given by the formulas

$$S_+ x = 0 \oplus \bigoplus_{k=1}^{\infty} U_k x_{k-1} \quad \text{and} \quad S_+^* x = \bigoplus_{k=0}^{\infty} U_{k+1}^* x_{k+1}$$

for every $x = \bigoplus_{k=0}^{\infty} x_k$ in $\mathcal{H} = \bigoplus_{k=0}^{\infty} \mathcal{H}_k$, where 0 is the origin of $\mathcal{H}_0$ and U_{k+1} is any unitary transformation of $\mathcal{H}_k$ onto $\mathcal{H}_{k+1}$ so that $S_+|_{\mathcal{H}_k} = U_{k+1}$, for each $k \geq 0$. These are identified with the infinite matrices

$$S_+ = \begin{pmatrix} O & & & \\ U_1 & O & & \\ & U_2 & O & \\ & & U_3 & O \\ & & & & \ddots \end{pmatrix} \quad \text{and} \quad S_+^* = \begin{pmatrix} O & U_1^* & & \\ & O & U_2^* & \\ & & O & U_3^* \\ & & & O \\ & & & & \ddots \end{pmatrix}$$

of transformations where every entry below (above) the main block diagonal in the matrix of S_+ (S_+^*) is unitary and the remaining entries are all null.

In particular, if $\mathcal{H}$ is an infinite-dimensional *separable* Hilbert space; that is, if the Hilbert space $\mathcal{H}$ has a countably infinite orthonormal basis, say $\{e_k\}_{k=0}^{\infty}$, then set $\mathcal{H}_k = \operatorname{span}\{e_k\}$ for each $k \geq 0$ so that $\dim \mathcal{H}_k = 1$ for all $k \geq 0$. It is easy to verify that S_+ is a unilateral shift of multiplicity 1 on $\mathcal{H}$ if and only if it shifts some orthonormal basis $\{e_k\}_{k=0}^{\infty}$ for $\mathcal{H}$; that is, if and only if $S_+ e_k = e_{k+1}$ for every $k \geq 0$.

Problem 5.1. Let S_+ be a unilateral shift on a Hilbert space. Show that

(a) S_+^* is a strongly stable coisometry,

(b) $S_+^n \xrightarrow{w} O$ but $\{S_+^n\}$ does not converge strongly.

Thus S_+ is an isometry and $S_+^{*n} \xrightarrow{s} O$ so that a backward unilateral shift S_+^* is a strongly stable coisometry. A unilateral shift in fact is the classical example of an isometry that is not a coisometry (i.e., of a nonunitary isometry or, equivalently, of a nonsurjective isometry). But what is more important is that this properly (being a strongly stable coisometry) actually characterizes backward unilateral shifts. Are the backward unilateral shifts the only strongly stable coisometries? Yes, they are. In fact, *an operator is a strongly stable coisometry if and only if it is the adjoint of a unilateral shift* (see e.g., [31, pp. 87,88]).

Let α (any nonzero cardinal number) be the multiplicity of a unilateral shift S_+. Let $\mathcal{K}_0$ be an arbitrary Hilbert space unitarily equivalent to $\mathcal{H}_0$ (for instance, $\mathcal{H}_0$ itself) so that $\dim \mathcal{H}_k = \dim \mathcal{K}_0 = \alpha$ for every $k \geq 0$. Let $U_0 \colon \mathcal{K}_0 \to \mathcal{H}_0$ be any unitary transformation of $\mathcal{K}_0$ onto $\mathcal{H}_0$. Now consider the Hilbert space $\ell_+^2(\mathcal{K}_0) = \bigoplus_{k=0}^{\infty} \mathcal{K}_0$ — the direct sum of countably infinite copies of $\mathcal{K}_0$. Since composition and direct sum of unitary transformations are again unitary, it follows that $U = \bigoplus_{k=0}^{\infty} U_k \cdots U_0 \colon \ell_+^2(\mathcal{K}_0) \to \mathcal{H}$ is unitary. Observe that

$$U^* S_+ U = \begin{pmatrix} O & & & \\ I & O & & \\ & I & O & \\ & & I & O \\ & & & \ddots \end{pmatrix} \quad \text{and} \quad U^* S_+^* U = \begin{pmatrix} O & I & & \\ & O & I & \\ & & O & I \\ & & & O \\ & & & \ddots \end{pmatrix},$$

in $\mathcal{B}[\ell_+^2(\mathcal{K}_0)]$. Thus S_+ is unitarily equivalent to $U^* S_+ U$, which is clearly a unilateral shift of multiplicity α acting on $\ell_+^2(\mathcal{K}_0)$. This is called the *canonical unilateral shift of multiplicity α*. If $\dim \mathcal{K}_0 = 1$ (i.e., if $\mathcal{K}_0 = \mathbb{C}$ in case of a complex space), then we get the canonical unilateral shift of multiplicity 1 acting on $\ell_+^2 = \ell_+^2(\mathbb{C})$, which is precisely the operator on ℓ_+^2 that shifts the canonical orthonormal basis for ℓ_+^2.

An operator S acting on a Hilbert space $\mathcal{H}$ is a *bilateral shift* if there exists an infinite family $\{\mathcal{H}_k\}_{k=-\infty}^{\infty}$ of nonzero pairwise orthogonal subspaces of $\mathcal{H}$ such that $\mathcal{H} = \bigoplus_{k=-\infty}^{\infty} \mathcal{H}_k$ (i.e., the orthogonal family $\{\mathcal{H}_k\}_{k=-\infty}^{\infty}$ spans $\mathcal{H}$) and S maps each $\mathcal{H}_k$ isometrically onto $\mathcal{H}_{k+1}$. As in the case of a unilateral shift, the above definition ensures that $S|_{\mathcal{H}_k} : \mathcal{H}_k \to \mathcal{H}_{k+1}$ is a surjective isometry so that the subspaces $\mathcal{H}_k$ of $\mathcal{H}$ are all unitarily equivalent and their common dimension is the *multiplicity* of S. The adjoint $S^* \in \mathcal{B}[\mathcal{H}]$ of $S \in \mathcal{B}[\mathcal{H}]$ is referred to as a *backward bilateral shift*. Writing $\bigoplus_{k=-\infty}^{\infty} x_k$ for $\{x_k\}_{k=-\infty}^{\infty}$ in $\bigoplus_{k=-\infty}^{\infty} \mathcal{H}_k$, it follows that the operators S and S^* are given by the formulas

$$Sx = \bigoplus_{k=-\infty}^{\infty} U_k x_{k-1} \quad \text{and} \quad S^*x = \bigoplus_{k=\infty}^{\infty} U_{k+1}^* x_{k+1}$$

for every $x = \bigoplus_{k=-\infty}^{\infty} x_k$ in $\mathcal{H} = \bigoplus_{k=-\infty}^{\infty} \mathcal{H}_k$, where $\{U_k\}_{k=-\infty}^{\infty}$ is any family of unitary transformations $U_{k+1} : \mathcal{H}_k \to \mathcal{H}_{k+1}$ so that $S|_{\mathcal{H}_k} = U_{k+1}$ for each k. These are identified with the following (doubly) infinite matrices of transformations (the inner parenthesis indicates the zero-zero entry).

$$S = \begin{pmatrix} \ddots & & & & \\ & O & & & \\ & U_{-1} & O & & \\ & & U_0 & (O) & \\ & & & U_1 & O \\ & & & & \ddots \end{pmatrix} \quad \text{and} \quad S^* = \begin{pmatrix} \ddots & & & & \\ & O & U_{-1}^* & & \\ & & O & U_0^* & \\ & & & (O) & U_1^* \\ & & & & O \\ & & & & \ddots \end{pmatrix}.$$

Again, every entry below (above) the main block diagonal in the matrix of S (S^*) is unitary and the remaining entries are all null. In particular, if $\mathcal{H}$ is an infinite-dimensional *separable* Hilbert space, then take any orthonormal basis $\{e_k\}_{k=-\infty}^{\infty}$ for $\mathcal{H}$ and set $\mathcal{H}_k = \text{span}\{e_k\}$ so that $\dim \mathcal{H}_k = 1$ for each integer $k \in \mathbb{Z}$. It is also easy to verify that S is a bilateral shift of multiplicity 1 on $\mathcal{H}$ if and only if it shifts some orthonormal basis $\{e_k\}_{k=-\infty}^{\infty}$ for $\mathcal{H}$; that is, if and only if $S_+ e_k = e_{k+1}$ for every $k \in \mathbb{Z}$. The same argument employed for unilateral shifts shows that S is unitarily equivalent to S_0, the *canonical bilateral shift of multiplicity* α acting on $\ell^2(\mathcal{K}_0) = \bigoplus_{k=-\infty}^{\infty} \mathcal{K}_0$,

$$S_0 = \begin{pmatrix} \ddots & & & & \\ & O & & & \\ & I & O & & \\ & & I & (O) & \\ & & & I & O \\ & & & & \ddots \end{pmatrix} \quad \text{so that} \quad S_0^* = \begin{pmatrix} \ddots & & & & \\ & O & I & & \\ & & O & I & \\ & & & (O) & I \\ & & & & O \\ & & & & \ddots \end{pmatrix},$$

in $\mathcal{B}[\ell^2(\mathcal{K}_0)]$, where $\mathcal{K}_0$ is any Hilbert space such that $\dim \mathcal{K}_0 = \alpha$. When $\dim \mathcal{K}_0 = 1$ (i.e., if $\mathcal{K}_0 = \mathbb{C}$ in case of a complex space) we get the canonical bilateral shift of multiplicity 1 acting on $\ell^2 = \ell^2(\mathbb{C})$, which is precisely the operator on ℓ^2 that shifts the canonical orthonormal basis for ℓ^2.

Problem 5.2. Let S be a bilateral shift on a Hilbert space. Show that

(a) S is a weakly stable unitary operator,

(b) $\{S^n\}$ and $\{S^{*n}\}$ do not converge strongly.

Besides being isometries, bilateral shifts are normal too (i.e., they are unitary operators). That makes a big difference between unilateral and bilateral shifts (unilateral shifts are not normal — they are prototypes of nonunitary isometries). Moreover, bilateral shifts are weakly stable unitary operators. Are they (and the direct summands of them) the only ones?

Question: Is it true that a weakly stable unitary operator must be a bilateral shift or a direct summand of a bilateral shift?

Problem 5.3. Show that two shifts (either both unilateral or both bilateral) are unitarily equivalent if and only if they have the same multiplicity.

Problem 5.4. If an operator is unitarily equivalent to a unilateral (bilateral) shift, then it is a unilateral (bilateral) shift of the same multiplicity.

Problem 5.5. Every unilateral shift of multiplicity greater than 1 is reducible.

Problem 5.6. Show that a unilateral shift of multiplicity 1 is irreducible.

Consider the Hilbert space $L^2(\Gamma)$ of all (equivalence classes of) square-summable complex functions defined on the unit circle Γ (about the origin of the complex plane $\mathbb{C}$) with respect to normalized Lebesgue measure on Γ (i.e., $\int_\Gamma dz = 1$). For each integer $k \in \mathbb{Z}$ put $e_k(z) = z^k$ for every $z \in \Gamma$. (Recall that, in this context, "equality" is interpreted almost everywhere in Γ.) Each e_k lies in $L^2(\Gamma)$. Actually, it is well known that $\{e_k\}_{k \in \mathbb{Z}}$ is an orthonormal basis for $L^2(\Gamma)$. Consider the mapping $U: L^2(\Gamma) \to L^2(\Gamma)$ that assigns to each function f in $L^2(\Gamma)$ the function Uf in $L^2(\Gamma)$ as follows.

$$(Uf)(z) = zf(z) \quad \text{for every} \quad z \in \Gamma.$$

It is readily verified that Uf, in fact, lies in $L^2(\Gamma)$ for every f in $L^2(\Gamma)$ and also that U is linear. Indeed, as we shall see below, $U \in \mathcal{B}[L^2(\Gamma)]$.

Problem 5.7. Show that U is a bilateral shift of multiplicity 1 on $L^2(\Gamma)$ that shifts the orthonormal basis $\{e_k\}_{k=-\infty}^{\infty}$.

Problem 5.8. Prove the Riemann–Lebesgue Lemma: If $f \in L^2(\Gamma)$, then

$$\int_\Gamma z^k f(z)\, dz \to 0 \quad \text{as} \quad k \to \pm\infty.$$

Remark: Shifts play a crucial role in operator theory, either bilateral or unilateral. It is clear, by their very definition, that they are operators on infinite-dimensional spaces. It is worth noticing that there are in current literature some different but equivalent definitions of them. Actually, shifts can be thought of as prototypes of infinite-dimensional operators (operators with an infinite-dimensional range). A bilateral shift is unitary, but unilateral shifts certainly are the most understood nonnormal operators. The problems in this chapter, although enough for further applications in this book, represent just a rather tiny *apéritif* on the subject. The interested reader is referred to [2], [9], [18], [20], [22], [31], [43] and [47].

Take a bounded scalar-valued sequence, say $\{\alpha_k\}_{k=0}^\infty$ in ℓ^∞. Consider the operator $T_+ = S_+ D$ in $\mathcal{B}[\ell_+^2]$, where $D = \mathrm{diag}(\{\alpha_k\}_{k=0}^\infty)$ is a diagonal on ℓ_+^2 and S_+ is the canonical unilateral shift (of multiplicity 1) on ℓ_+^2. Thus T_+ in $\mathcal{B}[\ell_+^2]$ is given by the formula $T_+ x = (0, \alpha_0 \xi_0, \alpha_1 \xi_1, \alpha_2 \xi_2, \dots)$ for every vector $x = (\xi_0, \xi_1, \xi_2, \dots)$ in ℓ_+^2. This is a *unilateral weighted shift* with weight sequence $\{\alpha_k\}_{k=0}^\infty$, usually denoted by $T_+ = \mathrm{shift}(\{\alpha_k\}_{k=0}^\infty) = \mathrm{shift}(\alpha_0, \alpha_1, \alpha_2, \dots)$, which can be represented as an infinite matrix

$$T_+ = \begin{pmatrix} 0 & & & \\ \alpha_0 & 0 & & \\ & \alpha_1 & 0 & \\ & & \alpha_2 & 0 \\ & & & & \ddots \end{pmatrix},$$

where every entry not below the main diagonal is null.

Problem 5.9. Let $T_+ = \mathrm{shift}(\{\alpha_k\}_{k=0}^\infty)$ be a unilateral weighted shift. Verify the following elementary properties.

(a) T_+ is injective if and only if $\alpha_k \neq 0$ for every k.

(b) $\|T_+\| = \sup_k |\alpha_k|$.

Problem 5.10. Let $T_+ = S_+ D$ be a unilateral weighted shift with weight sequence $\{\alpha_k\}_{k=0}^\infty$. Consider a scalar-valued sequence $\{\beta_k\}_{k=0}^\infty$ recursively defined as follows. Set $\beta_0 = 1$ and, for each $k \geq 0$,

$$\beta_{k+1} = \begin{cases} 1, & \alpha_k = 0, \\ \dfrac{\alpha_k}{|\alpha_k|} \beta_k, & \alpha_k \neq 0. \end{cases}$$

(a) Show that the diagonal $U = \mathrm{diag}(\{\beta_k\}_{k=0}^\infty)$ in $\mathcal{B}[\ell_+^2]$ is unitary.

(b) Put $|D| = (D^*D)^{\frac{1}{2}} = \mathrm{diag}(\{|\alpha_k|\}_{k=0}^\infty)$, a nonnegative diagonal on ℓ_+^2, and show that

$$S_+|D| = U^*S_+DU.$$

The above problem says that any unilateral weighted shift $T_+ = S_+D$ (with an arbitrary bounded complex weight sequence $\{\alpha_k\}_{k=0}^\infty$) is unitarily equivalent to the unilateral weighted shift $S_+|D|$ with nonnegative weight sequence $\{|\alpha_k|\}_{k=0}^\infty$. Thus, as far as unitary invariants are concerned, there is no loss of generality in considering unilateral weighted shifts with nonnegative weight sequences. For a comprehensive investigation on weighted shifts the reader is referred to [50].

Solutions

Solution 5.1. Let S_+ be a unilateral shift on $\mathcal{H} = \bigoplus_{k=0}^\infty \mathcal{H}_k$.

(a) Take an arbitrary vector $x = \bigoplus_{k=0}^\infty x_k$ in $\mathcal{H} = \bigoplus_{k=0}^\infty \mathcal{H}_k$ and note that $\|S_+x\|^2 = \sum_{k=1}^\infty \|U_k x_{k-1}\|^2 = \sum_{k=0}^\infty \|x_k\|^2 = \|x\|^2$. Hence S_+ is an isometry. This can also be verified by observing that $S_+^*S_+ = I$. Equivalently,

$$S_+^{*n}S_+^n = I \quad \text{for every} \quad n \geq 1,$$

which is immediately obtained by induction from the above representations of S_+ and S_+^* as operator matrices. As a matter of fact, since $S_+^{*n}x = \bigoplus_{k=0}^\infty U_{k+1}^* \cdots U_{k+n}^* x_{k+n}$ for each $n \geq 1$ (by induction again), it follows that $\|S_+^{*n}x\|^2 = \sum_{k=0}^\infty \|U_{k+1}^* \cdots U_{k+n}^* x_{k+n}\|^2 = \sum_{k=n}^\infty \|x_k\|^2$ (composition of unitary transformations is again unitary), and therefore S_+^* is strongly stable: $\|S_+^{*n}x\|^2 \to 0$ as $n \to \infty$ for every $x \in \mathcal{H}$.

(b) Recall that strong convergence implies weak convergence to the same limit, and also that an operator is weakly stable if and only if its adjoint is weakly stable. Thus it follows by (a) that

$$S_+^{*n} \xrightarrow{s} O \implies S_+^{*n} \xrightarrow{w} O \iff S_+^n \xrightarrow{w} O.$$

If $\{S_+^n\}$ converges strongly, then $S_+^n \xrightarrow{s} O$ because $S_+^n \xrightarrow{w} O$. But assertion (a) also says that S_+ is an isometry; that is, $\|S_+^n x\| = \|x\|$ for every x in $\mathcal{H}$, for all $n \geq 1$, and hence $S_+^n \not\xrightarrow{s} O$. Outcome: $\{S_+^n\}$ does not converge strongly (to any operator).

Solution 5.2. Let S be a bilateral shift on $\mathcal{H} = \bigoplus_{k=-\infty}^\infty \mathcal{H}_k$.

(a) Consider the canonical bilateral shift S_0 on $\ell^2(\mathcal{H}_0) = \bigoplus_{k=-\infty}^\infty \mathcal{H}_0$. Take an arbitrary pair of vectors $x = \bigoplus_{k=-\infty}^\infty x_k$ and $y = \bigoplus_{k=-\infty}^\infty y_k$

in $\ell^2(\mathcal{H}_0)$, and take an arbitrary integer $n \geq 1$. A trivial induction shows that $S_0^n x = \bigoplus_{k=-\infty}^{\infty} x_{k-n}$. If n is even put $\tilde{n} = \bar{n} = \frac{n}{2}$, if n is odd put $\tilde{n} = \frac{n-1}{2}$ and $\bar{n} = \frac{n+1}{2}$; so that $\tilde{n} + \bar{n} = n$. Thus

$$\langle S_0^n x ; y \rangle = \sum_{k=-\infty}^{\infty} \langle x_{k-n} ; y_k \rangle = \sum_{k=-\infty}^{-\tilde{n}-1} \langle x_k ; y_{k+n} \rangle + \sum_{k=-\tilde{n}}^{\infty} \langle x_k ; y_{k+n} \rangle.$$

Using the Schwarz inequality in $\mathcal{H}_0$ and in ℓ^2 we get

$$\sum_{k=-\infty}^{-\tilde{n}-1} |\langle x_k ; y_{k+n} \rangle| \leq \sum_{k=-\infty}^{-\tilde{n}-1} \|x_k\| \|y_{k+n}\|$$

$$\leq \left(\sum_{k=-\infty}^{-\tilde{n}-1} \|x_k\|^2 \right)^{\frac{1}{2}} \left(\sum_{k=-\infty}^{-\tilde{n}-1} \|y_{k+n}\|^2 \right)^{\frac{1}{2}}$$

$$\leq \left(\sum_{k=-\infty}^{-\tilde{n}-1} \|x_k\|^2 \right)^{\frac{1}{2}} \|y\| \to 0$$

and, similarly,

$$\sum_{k=-\tilde{n}}^{\infty} |\langle x_k ; y_{k+n} \rangle| \leq \|x\| \left(\sum_{k=-\tilde{n}}^{\infty} \|y_{k+n}\|^2 \right)^{\frac{1}{2}} = \|x\| \left(\sum_{j=\bar{n}}^{\infty} \|y_j\|^2 \right)^{\frac{1}{2}} \to 0,$$

as $n \to \infty$ because x and y lie in $\ell^2(\mathcal{H}_0)$ (so that $\sum_{k=-\infty}^{\infty} \|x_k\|^2 < \infty$ and $\sum_{j=-\infty}^{\infty} \|y_j\|^2 < \infty$) and both $\tilde{n}$ and $\bar{n}$ are strictly increasing positive unbounded (coercive) functions of n. Hence

$$|\langle S_0^n x ; y \rangle| \leq \sum_{k=-\infty}^{-\tilde{n}-1} |\langle x_k ; y_{k+n} \rangle| + \sum_{k=-\tilde{n}}^{\infty} |\langle x_k ; y_{k+n} \rangle| \to 0$$

as $n \to \infty$ so that $S_0^n \xrightarrow{w} O$. Since S_0 is unitarily equivalent to S, it follows that $S^n = (US_0 U^*)^n = US_0^n U^*$ for each integer $n \geq 1$, for some unitary $U : \ell^2(\mathcal{H}_0) \to \mathcal{H}$, and therefore $S^n \xrightarrow{w} O$; that is, S is weakly stable. Moreover, $S^* S = SS^* = I$ so that S is unitary.

(b) S is weakly stable. Thus $S^n \xrightarrow{w} O$ so that $S^{*n} \xrightarrow{w} O$. Since strong stability implies weak stability to the same limit, it follows that, if $\{S^n\}$ (or $\{S^{*n}\}$) converges strongly, then it must converge to O. However, S is unitary, which means that S is an isometry and a coisometry so that $\|S^n x\| = \|S^{*n} x\| = \|x\|$ for every x in $\mathcal{H}$, for all $n \geq 1$, and hence $\{S^n\}$ and $\{S^{*n}\}$ do not converge strongly to O. Therefore, both $\{S^n\}$ and $\{S^{*n}\}$ do not converge strongly (to any operator).

Solution 5.3. Recall that unitary equivalence is, of course, transitive. Two shifts S_1 and S_2 (both unilateral or both bilateral) are unitarily equivalent if and only if they are unitarily equivalently to the canonical (unilateral or bilateral, respectively) shift of multiplicity, say, α. Hence S_1 and S_2 have the same multiplicity α. Conversely, if two shifts (both unilateral or both bilateral) S_1 and S_2 have the same multiplicity α, then they are unitarily equivalent to the canonical (either unilateral or bilateral, respectively) shift of multiplicity α, and hence S_1 and S_2 are unitarily equivalent.

Solution 5.4. Let S be a shift (either unilateral or bilateral) of multiplicity α acting on a Hilbert space $\mathcal{H}$, and let $\{\mathcal{H}_k\}$ be the underlying countably infinite orthogonal family of unitarily equivalent subspaces of $\mathcal{H} = \bigoplus_k \mathcal{H}_k$ so that $\dim \mathcal{H}_k = \alpha$ for all k. Take an operator T acting on a Hilbert space $\mathcal{K}$ and suppose there exists a unitary transformation $U : \mathcal{K} \to \mathcal{H}$ such that

$$T = U^* S U.$$

For each k set $\mathcal{K}_k = U^*(\mathcal{H}_k)$ so that $\mathcal{H}_k = U(\mathcal{K}_k)$. Since U is unitary, it follows that $\{\mathcal{K}_k\}$ is an orthogonal family (recall: $\{\mathcal{H}_k\}$ is orthogonal and U preserves inner product) of subspaces of $\mathcal{K}$ (reason: range of isometry is closed). Moreover, $\{\mathcal{K}_k\}$ spans $\mathcal{K}$; that is, $\mathcal{K} = \bigoplus_k \mathcal{K}_k$. (Proof: if x lies in $(\bigoplus_k \mathcal{K}_k)^\perp$, then $x \perp \mathcal{K}_k = U^*(\mathcal{H}_k)$ so that $Ux \perp \mathcal{H}_k$ for all k, and hence Ux lies in $(\bigoplus_k \mathcal{H}_k)^\perp$, which is the null subspace $\{0\}$ because $\mathcal{H} = \bigoplus_k \mathcal{H}_k$, and therefore $Ux = 0$; that, is $x = 0$.) Furthermore, since $S_+(\mathcal{H}_k) = \mathcal{H}_{k+1}$,

$$T(\mathcal{K}_k) = U^* S U(\mathcal{K}_k) = U^* S(\mathcal{H}_k) = U^*(\mathcal{H}_{k+1}) = \mathcal{K}_{k+1},$$

for each k. Finally note that $T^*T = U^* S^* S U = I$. Thus T is an isometry, and so is any restriction of it (in particular, $T|_{\mathcal{K}_k}$ is an isometry for every k). Summing up: T is a shift (either unilateral or bilateral) with the same multiplicity of S once $\dim \mathcal{K}_k = \dim U^*(\mathcal{H}_k) = \dim \mathcal{H}_k = \alpha$ for all k.

Solution 5.5. Let $\mathcal{H}$ be any Hilbert space of dimension greater than 1 and let I be the identity on $\mathcal{H}$. Since $\dim \mathcal{H} > 1$, there exists a nontrivial orthogonal projection E on $\mathcal{H}$. Consider the operators S_+ and P,

$$S_+ = \begin{pmatrix} O & & & \\ I & O & & \\ & I & O & \\ & & I & O \\ & & & & \ddots \end{pmatrix} \quad \text{and} \quad P = \begin{pmatrix} E & & & \\ & E & & \\ & & E & \\ & & & E \\ & & & & \ddots \end{pmatrix},$$

in $\mathcal{B}[\ell_+^2(\mathcal{H})]$, where S_+ is the canonical unilateral shift of multiplicity $\dim \mathcal{H}$ and $P = \bigoplus_{k=0}^\infty E$ is a block diagonal, both acting on $\ell_+^2(\mathcal{H}) = \bigoplus_{k=0}^\infty \mathcal{H}$. E is a nontrivial orthogonal projection, and so is P. In fact, $O \neq E \neq I$ and $E^* = E = E^2$ implies $O \neq P \neq I$ and $P^* = P = P^2$. It is readily verified that $PS_+ = S_+P$ so that the canonical unilateral shift S_+ commutes with the nontrivial orthogonal projection P. Thus S_+ is reducible by Problem 4.6, and so is *every* unilateral shift of multiplicity greater than 1 according to Problems 4.9 and 5.3.

Solution 5.6. Let S_+ be the canonical unilateral shift of multiplicity 1 acting on $\ell_+^2 = \ell_+^2(\mathbb{C})$,

$$S_+ = \begin{pmatrix} 0 & & & \\ 1 & 0 & & \\ & 1 & 0 & \\ & & 1 & 0 \\ & & & & \ddots \end{pmatrix} \quad \text{and} \quad S_+^* = \begin{pmatrix} 0 & 1 & & \\ & 0 & 1 & \\ & & 0 & 1 \\ & & & 0 \\ & & & & \ddots \end{pmatrix}.$$

Take the canonical orthonormal basis $\{e_k\}_{k=0}^\infty$ for ℓ_+^2 and, for each $k \geq 0$, consider the orthogonal projection

$$P_k = \operatorname{diag}(e_k) = \operatorname{diag}(0, \dots, 0, 1, 0, \dots)$$

in $\mathcal{B}[\ell_+^2]$ (just one nonzero entry, equal to 1, after k zeros). Observe that

$$S_+^k(I - S_+S_+^*)S_+^{*k} = P_k \quad \text{for every} \quad k \geq 0,$$

which is readily verified by induction once $S_+P_kS_+^* = P_{k+1}$ for each $k \geq 0$. Suppose S_+ is reducible; that is, suppose there exists a *nontrivial* subspace of ℓ_+^2, say $\mathcal{M}$, which is invariant for both S_+ and S_+^*. Since $\mathcal{M}$ is a nonzero linear manifold of ℓ_+^2, there exists a nonzero vector $x = \{\xi_k\}_{k=0}^\infty$ in $\mathcal{M}$ such that $\xi_n = 1$ for some integer $n \geq 0$. Since $\mathcal{M}$ is invariant for both S_+ and S_+^*, it follows that $S_+^n(I - S_+S_+^*)S_+^{*n}x$ lies in $\mathcal{M}$, and hence $e_n = P_nx$ lies in $\mathcal{M}$ according to the above identity. Then $e_{n+k} = S_+^k e_n \in \mathcal{M}$ for every $k \geq 1$ ($\mathcal{M}$ is S_+-invariant) and $e_{n-k} = S_+^{*k}e_n \in \mathcal{M}$ for each $0 \leq k \leq n$ ($\mathcal{M}$ is S_+^*-invariant) so that $e_k \in \mathcal{M}$ for every $k \geq 0$. Hence $\operatorname{span}\{e_k\}_{k=0}^\infty \subseteq \mathcal{M}$. But $\{e_k\}_{k=0}^\infty$ is an orthonormal basis for the Hilbert space ℓ_+^2 and $\mathcal{M}$ is closed in ℓ_+^2. Thus $\ell_+^2 = (\operatorname{span}\{e_k\}_{k=0}^\infty)^- \subseteq \mathcal{M}^- = \mathcal{M}$ so that $\mathcal{M} = \ell_+^2$; a contradiction ($\mathcal{M}$ is a proper subspace of ℓ_+^2). Outcome: S_+ is irreducible. Therefore, according to Problems 4.9 and 5.3, we may conclude that *every* unilateral shift of multiplicity 1 is irreducible.

Solution 5.7. First note that U is an invertible isometry, thus a unitary operator in $\mathcal{B}[L^2(\Gamma)]$. Indeed, $\|Uf\|^2 = \int_\Gamma |zf(z)|^2 dz = \int_\Gamma |f(z)|^2 dz = \|f\|^2$ for every f in $L^2(\Gamma)$. Moreover, the mapping $U^{-1}: L^2(\Gamma) \to L^2(\Gamma)$ that assigns to each function f in $L^2(\Gamma)$ the function $U^{-1}f$ in $L^2(\Gamma)$ given by $(U^{-1}f)(z) = z^{-1}f(z)$ for every z in Γ is such that $U^{-1}U = UU^{-1} = I$. Finally, note that U shifts the orthonormal basis $\{e_k\}_{k=-\infty}^\infty$ for $L^2(\Gamma)$. In fact, $(Ue_k)(z) = ze_k(z) = z^{k+1} = e_{k+1}(z)$ for every z in Γ, for each $k \in \mathbb{Z}$. Thus U is a bilateral shift of multiplicity 1 on $L^2(\Gamma)$.

Solution 5.8. Let U be the bilateral shift of Problem 5.7. Take any f in $L^2(\Gamma)$. By induction we get $(U^k f)(z) = z^k f(z)$ and $(U^{-k} f)(z) = z^{-k} f(z)$ for each $k \geq 0$ and every $z \in \Gamma$. Thus $\langle U^k f ; 1 \rangle = \int_\Gamma z^k f(z)\, dz$ for every k in $\mathbb{Z}$, where $1(z) = 1$ for all z in Γ. But $U^k \xrightarrow{\ w\ } O$ and $U^{-k} \xrightarrow{\ w\ } O$ (Problem 5.2 with $U^* = U^{-1}$). Hence $\int_\Gamma z^k f(z)\, dz = \langle U^k f ; 1 \rangle \to 0$ as $k \to \pm\infty$.

Solution 5.9. Suppose $T_+ = S_+ D$, where S_+ is a canonical unilateral shift and $D = \mathrm{diag}(\{\alpha_k\}_{k=0}^\infty)$. Recall that T_+ is injective if and only if $T_+^* T_+$ is injective (see Problem 2.1(a)). But $T_+^* T_+ = D^* S_+^* S_+ D = D^* D$ because S_+ is an isometry (i.e., $S_+^* S_+ = I$). Since $D^* D = \mathrm{diag}(\{|\alpha_k|^2\}_{k=0}^\infty)$, it follows that $D^* D$ is injective (i.e., $\mathcal{N}(D^* D) = \{0\}$) if and only if $|\alpha_k| \neq 0$ for every k. Summing up:

(a) T_+ is injective if and only if $\alpha_k \neq 0$ for every k.

Let $\{e_k\}_{k=0}^\infty$ be the canonical basis for ℓ_+^2 so that $T_+ e_k = \alpha_k e_{k+1}$ for each k, and hence $\|T_+\| = \sup_{\|x\|=1} \|Tx\| \geq \|T e_k\| = |\alpha_k|$ for all k. On the other hand, note that $\|T_+\| \leq \|S_+\| \|D\| = \|D\| = \sup_k |\alpha_k|$ (cf. Problem 3.1(a) and recall that S_+ is an isometry and so $\|S_+\| = 1$). Therefore,

(b)
$$\|T_+\| = \sup_k |\alpha_k|.$$

Solution 5.10. Set $U = \mathrm{diag}(\{\beta_k\}_{k=0}^\infty)$ on the Hilbert space ℓ_+^2, where $\{\beta_k\}_{k=0}^\infty$ is recursively defined as follows. $\beta_0 = 1$ and, for each $k \geq 0$,

$$
\beta_{k+1} = \begin{cases} 1, & \alpha_k = 0, \\[2mm] \dfrac{\alpha_k}{|\alpha_k|} \beta_k, & \alpha_k \neq 0. \end{cases}
$$

(a) Since $|\beta_{k+1}| = |\beta_k|$ for every $k \geq 0$ and $\beta_0 = 1$, a trivial induction ensures that $|\beta_k| = 1$ for every $k \geq 0$. Hence $\|Ux\|^2 = \sum_{k=0}^\infty |\beta_k \xi_k|^2 = \|x\|^2$ for every $x = \{\xi_k\}_{k=0}^\infty \in \ell_+^2$ so that U is an operator (an isometry, actually) in $\mathcal{B}[\ell_+^2]$. Since $U^* = \mathrm{diag}(\{\overline{\beta_k}\}_{k=0}^\infty)$, it follows that $U^* U = UU^* = \mathrm{diag}(\{|\beta_k|^2\}_{k\geq 0}) = I$, the identity on ℓ_+^2, and so U is unitary.

(b) Note that $|\alpha_k| \beta_{k+1} = \alpha_k \beta_k$ for each $k \geq 0$. Hence

$$
\begin{pmatrix} 0 & & & \\ \beta_1|\alpha_0| & 0 & & \\ & \beta_2|\alpha_1| & 0 & \\ & & \beta_3|\alpha_2| & 0 \\ & & & & \ddots \end{pmatrix} = \begin{pmatrix} 0 & & & \\ \alpha_0\beta_0 & 0 & & \\ & \alpha_1\beta_1 & 0 & \\ & & \alpha_2\beta_2 & 0 \\ & & & & \ddots \end{pmatrix}
$$

so that $U S_+ |D| = S_+ D U$.

6

Decompositions

A *contraction* is an operator T on a normed space $\mathcal{X}$ such that $\|T\| \leq 1$. Equivalently, such that $\|Tx\| \leq \|x\|$ for every x in $\mathcal{X}$. If T is a contraction on a Hilbert space $\mathcal{H}$, then $\{T^{*n}T^n\}$ is a decreasing sequence of nonnegative contractions. In fact, take an arbitrary positive integer n. Since $T^{*n} = T^{n*}$ we get $T^{*n}T^n \geq O$ and $\|T^{*n}T^n\| \leq \|T^{*n}\|\|T^n\| \leq \|T\|^{2n} \leq 1$. Moreover, $\langle T^{*n+1}T^{n+1}x\,;x\rangle = \|T^{n+1}x\|^2 \leq \|T^n x\|^2 = \langle T^{*n}T^n x\,;x\rangle$ for every x in $\mathcal{H}$. Thus $\{T^{*n}T^n\}$ is a bounded monotone sequence of self-adjoint operators, and therefore it converges strongly (Problem 3.5). Summing up: if T is a contraction on a Hilbert space $\mathcal{H}$, then

$$T^{*n}T^n \xrightarrow{\ s\ } A,$$

for some operator A on $\mathcal{H}$; the strong limit of $\{T^{*n}T^n\}$.

Problem 6.1. Show that the strong limit A of $\{T^{*n}T^n\}$ has the following properties.

(a) $O \leq A \leq I$ (i.e., $A \in \mathcal{B}^+[\mathcal{H}]$ and $\|A\| \leq 1$).

(b) $\|T^n x\| \to \|A^{\frac{1}{2}} x\|$ for every $x \in \mathcal{H}$.

(c) $T^{*n}AT^n = A$ for every $n \geq 1$.

(d) $\|A^{\frac{1}{2}}T^n x\| = \|A^{\frac{1}{2}} x\|$ for every $x \in \mathcal{H}$ and every $n \geq 1$.

(e) $AT \neq O$ and $TA \neq O$ whenever $A \neq O$.

(f) $\|A\| = 1$ whenever $A \neq O$.

Moreover, show that $(I - A)T^n \xrightarrow{s} O$, and hence

(g) $\|AT^n x\| \to \|A^{\frac{1}{2}} x\|$ for every $x \in \mathcal{H}$.

Finally, verify that $(I - A^{\frac{1}{2}})T^n \xrightarrow{s} O$.

Problem 6.2. Consider the setup of Problem 6.1. Show that

(a) $A = O$ if and only if T is strongly stable,

(b) $A = I$ if and only if T is an isometry,

(c) $A = A^2$ if and only if A commutes with T.

Problem 6.3. Consider again the setup of Problem 6.1. Show that

(a) $\mathcal{N}(A) = \{x \in \mathcal{H}: T^n x \to 0\}$,

(b) $\mathcal{N}(I - A) = \{x \in \mathcal{H}: \|A^{\frac{1}{2}} x\| = \|x\|\}$
$= \{x \in \mathcal{H}: \|T^n x\| = \|x\| \text{ for every } n \geq 1\}$,

(c) $\mathcal{N}(A - A^2) = \{x \in \mathcal{H}: \|A^{\frac{1}{2}} x\| = \|Ax\|\}$
$= \{x \in \mathcal{H}: \|AT^n x\| = \|Ax\| \text{ for every } n \geq 1\}$
$= \{x \in \mathcal{H}: AT^n x = T^n Ax \text{ for every } n \geq 1\}$.

Note that each of the identities in Problem 6.3 supplies another proof for each assertion in Problem 6.2. Moreover, it is readily verified from Problem 6.3 that $\mathcal{N}(A)$, $\mathcal{N}(I - A)$ and $\mathcal{N}(A - A^2)$ are invariant subspaces for T. In fact, $\mathcal{N}(A)$ is hyperinvariant for T. Here is an invariant subspace decomposition for the invariant subspace $\mathcal{N}(A - A^2)$.

Problem 6.4. Show that $\mathcal{N}(A - A^2) = \mathcal{N}(A) \oplus \mathcal{N}(I - A)$.

Problem 6.5. Let T be a contraction on a Hilbert space, and let A be the strong limit of $\{T^{*n} T^n\}$ as in Problem 6.1. Show that there exists an isometry V in $\mathcal{B}[\mathcal{R}(A)^-]$ such that

$$A^{\frac{1}{2}} T = V A^{\frac{1}{2}}.$$

Thus, if $A \neq O$, then $A^{\frac{1}{2}}$ intertwines T to an isometry V on $\mathcal{R}(A)^- \neq \{0\}$.

Hint: Verify that the restriction of $A^{\frac{1}{2}}$ to $\mathcal{R}(A^{\frac{1}{2}})$ is injective. Now consider the map $V_0: \mathcal{R}(A^{\frac{1}{2}}) \to \mathcal{R}(A^{\frac{1}{2}})$ that assigns to each vector y in $\mathcal{R}(A^{\frac{1}{2}})$ the vector $V_0 y = A^{\frac{1}{2}} T x$ in $\mathcal{R}(A^{\frac{1}{2}})$, where x is the *unique* vector in $\mathcal{H}$ such that $y = A^{\frac{1}{2}} x$. Recall that $\mathcal{R}(A^{\frac{1}{2}})^- = \mathcal{R}(A)^-$ and extend V_0 by continuity (see e.g., [32, p. 237]) to $\mathcal{R}(A)^-$. This yields the isometry $V \in \mathcal{B}[\mathcal{R}(A)^-]$.

Such an isometry V induced by any contraction T has many interesting and useful further properties (see e.g., [16], [31, pp. 55–60] and [52, p. 40]).

Recall that T is a contraction if and only if T^* is. Thus let the nonnegative contraction A_* in $\mathcal{B}[\mathcal{H}]$ be the strong limit of $\{T^n T^{*n}\}$ and let V_* in $\mathcal{B}[\mathcal{R}(A_*)^-]$ be the isometry such that $A_*^{1/2} T^* = V_* A_*^{1/2}$. Clearly, each of the results in Problems 6.1 to 6.5 holds for A and V replaced with A_* and V_*, respectively, provided that T is replaced with T^*. Let us just mention that the isometries V and V_* (as well as the operators A and A_*), which we have associated with a contraction T, can be extended to a more general class of operators. Indeed, there are counterparts of them for power bounded operators (see [23] and [24]).

Problem 6.6. Show that if $A = A_*$, then $A = A^2$.

A contraction is *completely nonunitary* if it has no nonzero unitary direct summand; equivalently, if the restriction of it to any nonzero reducing subspace is not unitary. The next result is a central one about contractions on Hilbert spaces (see [52, p. 9]). It characterizes the largest reducing subspace on which a contraction is unitary, thus establishing that *every contraction on a Hilbert space has a unique decomposition into a direct* (orthogonal) *sum of a completely nonunitary contraction and a unitary operator.*

Nagy–Foiaş–Langer Decomposition. *Let T be a contraction acting on a Hilbert space $\mathcal{H}$ and set*

$$\mathcal{U} = \mathcal{N}(I - A) \cap \mathcal{N}(I - A_*).$$

$\mathcal{U}$ is a reducing subspace for T. Moreover, the decomposition

$$T = C \oplus U$$

on $\mathcal{H} = \mathcal{U}^\perp \oplus \mathcal{U}$ is such that $C = T|_{\mathcal{U}^\perp}$ is a completely nonunitary contraction and $U = T|_{\mathcal{U}}$ is unitary.

Problem 6.7. Prove the Nagy–Foiaş–Langer decomposition.

A Hilbert space operator is *pure* if it has no normal direct summand (i.e., if it is completely nonnormal). In particular, an isometry is pure if it has no normal isometry as a direct summand (since the restriction of an isometry to an invariant subspace is again an isometry). But a normal isometry is precisely a unitary operator. Thus pure isometry means completely nonunitary isometry. However, *an operator is a completely nonunitary isometry if and only if it is a unilateral shift.* This is a well-known result for operators on Hilbert spaces (see e.g., [31, p. 79]). Therefore, *pure isometry, completely nonunitary isometry* and *unilateral shift* are synonyms on a Hilbert space.

Since isometries are contractions, the Nagy–Foiaş–Langer decomposition holds for isometries in particular, now with C standing for a completely nonunitary isometry (a completely nonunitary restriction of an isometry to a reducing subspace), which means a unilateral shift. This leads to a classical decomposition for isometries, which chronologically precedes Nagy–Foiaş–Langer decomposition. *Every isometry on a Hilbert space is either a unilateral shift, a unitary operator, or a direct (orthogonal) sum of them.*

von Neumann–Wold Decomposition. *If T is an isometry on a Hilbert space $\mathcal{H}$, then $\mathcal{N}(I - A_*)$ is a reducing subspace for T. Moreover, the decomposition*

$$T = S_+ \oplus U$$

on $\mathcal{H} = \mathcal{N}(I - A_*)^\perp \oplus \mathcal{N}(I - A_*)$ is such that $S_+ = T|_{\mathcal{N}(I-A_*)^\perp}$ is a unilateral shift and $U = T|_{\mathcal{N}(I-A_*)}$ is unitary.

Problem 6.8. Apply the Nagy–Foiaş–Langer decomposition for contractions to prove the von Neumann–Wold decomposition for isometries.

If $A = O$, then T is strongly stable, and hence a completely nonunitary contraction. If $A = I$, then T is an isometry. We shall consider now a more general situation where A (or A_*) is required to be just a projection. The class of all contractions T such that A is a projection (that is, such that $\mathcal{N}(A - A^2) = \mathcal{H}$) was investigated in [12] and [38] and surveyed in [33]. It coincides with the class of all contractions T that commute with A; that is, $A = A^2$ if and only if $AT = TA$ (Problem 6.2(c)).

Problem 6.9. Let T be a contraction on a Hilbert space. Show that

(a) $A = A^2$ if and only if $T = G \oplus S_+ \oplus U$,

where G is a strongly stable contraction on $\mathcal{N}(A)$, S_+ is a unilateral shift on $\mathcal{N}(I - A) \cap \mathcal{N}(A_*)$, and U is a unitary operator on $\mathcal{N}(I - A) \cap \mathcal{N}(I - A_*)$;

(b) $A = A^2$ and $A_* = A_*^2$ if and only if $T = B \oplus S_- \oplus S_+ \oplus U$,

where B is a contraction on $\mathcal{N}(A) \cap \mathcal{N}(A_*)$ such that both B and B^* are strongly stable and S_- is a backward unilateral shift (i.e., the adjoint of a unilateral shift) on $\mathcal{N}(A) \cap \mathcal{N}(I - A_*)$;

(c) $A = A_*$ if and only if $T = B \oplus U$.

Another useful decomposition for contractions on a Hilbert space reads as follows (for a proof see e.g., [31, p. 103]).

Foguel Decomposition. *If T is a contraction on a Hilbert space, then*

$$T = Z \oplus U,$$

where Z is a weakly stable contraction and U is unitary.

Of course, as in the previous decompositions, it is understood that any of the above direct summands may be missing. Nagy–Foiaş–Langer decomposition is useful in reducing questions about contractions to completely nonunitary contractions. Foguel decomposition is also useful in reducing them to weakly stable contractions instead. Unlike Nagy–Foiaş–Langer decomposition, Foguel decomposition is not unique. Example: every bilateral shift is a weakly stable unitary operator (Problem 5.2), and so is any direct summand of it. Use the Foguel decomposition to prove the next result.

Problem 6.10. Every completely nonunitary contraction is weakly stable.

Now we borrow from [13] the following definition (as well as the final problem in this chapter; also see [35]). Let $\mathcal{H}$ and $\mathcal{K}$ be Hilbert spaces. A contraction $T \in \mathcal{B}[\mathcal{H}]$ has *property PF* (a short for Putnam–Fuglede) if, whenever the equation

$$XT^* = JX$$

holds for some isometry $J \in \mathcal{B}[\mathcal{K}]$ and some $X \in \mathcal{B}[\mathcal{H}, \mathcal{K}]$, then

$$XT = J^*X.$$

That is, a contraction T has property PF if either T^* is not intertwined to any isometry or, if T^* is intertwined to an isometry J, then the same transformation that intertwines T^* to J also intertwines T to the coisometry J^*. The next problems state basic facts about contractions with property PF.

Problem 6.11. Every isometry has property PF.

Problem 6.12. If a contraction T has a nonunitary coisometry as a direct summand, then T does not have property PF. In particular, if a coisometry has property PF, then it is unitary.

Problem 6.13. A contraction T on a Hilbert space $\mathcal{H}$ is strongly stable if and only if the unique solution X in $\mathcal{B}[\mathcal{H}, \mathcal{K}]$ to the equation $XT = JX$ for any isometry J on any Hilbert space $\mathcal{K}$ is the trivial $X = O$.

Corollary: If T is a strongly stable contraction (i.e., if $A = O$), then T^* has property PF. Indeed, if $T^n \xrightarrow{s} O$, then $T = T^{**}$ is not intertwined to any isometry (Problem 6.13), and hence T^* has property PF.

The class of contractions T for which $A = A^2$ includes the class of contractions T such that T^* has property PF (Problem 6.14 below). Therefore, the above corollary, Problem 6.11, and the next problem ensure that

$$A = O \text{ or } A = I \implies T^* \text{ has property PF} \implies A = A^2.$$

Problem 6.14. If a contraction T has property PF, then $A_* = A_*^2$.

The converse fails: if T is a nonunitary coisometry (e.g., a backward unilateral shift), then $A_* = I$ but T does not have property PF (Problem 6.12). Now apply the above result and the decompositions of Problems 6.7, 6.8 and 6.9 to prove the following assertion.

Problem 6.15. The completely nonunitary direct summand of a contraction T is strongly stable if and only if T^* has property PF.

Solutions

Solution 6.1. Let A be the strong limit of $\{T^{*n}T^n\}$ and consider the assertions (a)–(g) in the problem statement.

(a) A is nonnegative because it is the strong limit of a nonnegative sequence (Problem 3.8: $\mathcal{B}^+[\mathcal{H}]$ is weakly, thus strongly, closed in $\mathcal{B}[\mathcal{H}]$). A is a contraction because it is the strong limit of a sequence of contractions. Indeed, $\|Ax\| = \lim \|T^{*n}T^n x\| \leq \|x\|$ for every $x \in \mathcal{H}$ since norm is continuous. Equivalently, $O \leq A \leq I$ (Problem 2.3).

(b) Strong convergence implies weak convergence. Since A is nonnegative, let $A^{\frac{1}{2}}$ be the nonnegative square root of A. Thus, by continuity of the inner product, $\|T^n x\|^2 = \langle T^{*n}T^n x ; x \rangle \to \langle Ax; x \rangle = \|A^{\frac{1}{2}}x\|^2$ for every $x \in \mathcal{H}$.

(c) First note that $T^{*k+n}T^{k+n} = T^{*n}T^{*k}T^k T^n$ for every $k, n \geq 1$. But, for each $n \geq 1$, $T^{*k+n}T^{k+n} \xrightarrow{s} A$ and $T^{*n}T^{*k}T^k T^n \xrightarrow{s} T^{*n}AT^n$ as $k \to \infty$ because $T^{*k}T^k \xrightarrow{s} A$. The identity in (c) then follows by uniqueness of the limit. That is, $T^{*n}AT^n = A$ for every integer $n \geq 1$.

(d) If (c) holds, then $\|A^{\frac{1}{2}}T^n x\|^2 = \langle T^{*n}AT^n x ; x \rangle = \langle Ax ; x \rangle = \|A^{\frac{1}{2}}x\|^2$ for every $x \in \mathcal{H}$ and every $n \geq 1$.

(e) Note from (c) that if $AT = O$, then $A = O$. Moreover, (c) also ensures that $TA = O$ implies $A^2 = O$. Since the square of any self-adjoint operator is nonnegative, and since the nonnegative square root of a nonnegative operator is unique, it follows that $A = O$.

(f) According to (b) and (d),

$$\|A^{\frac{1}{2}}x\| = \|A^{\frac{1}{2}}T^n x\| \leq \|A^{\frac{1}{2}}\|\|T^n x\| \to \|A^{\frac{1}{2}}\|\|A^{\frac{1}{2}}x\| \leq \|A^{\frac{1}{2}}\|^2\|x\|$$

for every $x \in \mathcal{H}$, and hence $\|A^{\frac{1}{2}}\| \leq \|A^{\frac{1}{2}}\|^2$. Since $\|A^{\frac{1}{2}}\|^2 = \|A\| \leq 1$, it follows that either $\|A\| = 1$ or $\|A\| = 0$.

Now observe from (c) that $T^{*n}(I - A)T^n = T^{*n}T^n - A \xrightarrow{s} O$. Therefore, since (a) says that $O \leq I - A \leq I$, so that $I - A$ is a nonnegative contraction and so is $(I - A)^{\frac{1}{2}}$, and since strong convergence implies weak convergence,

$$\|(I - A)T^n x\|^2 \leq \|(I - A)^{\frac{1}{2}}T^n x\|^2 = \langle T^{*n}(I - A)T^n x; x\rangle \to 0$$

for every $x \in \mathcal{H}$. Thus $(I - A)T^n \xrightarrow{s} O$.

(g) Moreover, since A is a contraction, it follows that $\|AT^n x\| \leq \|T^n x\|$, and hence $0 \leq \|T^n x\| - \|AT^n x\| \leq \|(I - A)T^n x\|$, for all $x \in \mathcal{H}$ and every $n \geq 1$. Then $\|T^n x\| - \|AT^n x\| \to 0$ so that $\|AT^n x\| \to \|A^{\frac{1}{2}} x\|$ for every $x \in \mathcal{H}$ according to (b).

Finally, observe that $(I - A)T^n \xrightarrow{s} O$ implies $(I - A^{\frac{1}{2}})T^n \xrightarrow{s} O$. Indeed, $I + A^{\frac{1}{2}}$ has a bounded inverse (reason: $I + A^{\frac{1}{2}} \succ O$ because $A^{\frac{1}{2}} \geq O$), and therefore (since $I - A = (I + A^{\frac{1}{2}})(I - A^{\frac{1}{2}})$ and $(I + A^{\frac{1}{2}})^{-1}$ is bounded)

$$\|(I - A^{\frac{1}{2}})T^n x\| = \|(I + A^{\frac{1}{2}})^{-1}(I - A)T^n\| \leq \|(I + A^{\frac{1}{2}})^{-1}\| \|(I - A)T^n x\|$$

for every $x \in \mathcal{H}$. Hence $\|(I - A^{\frac{1}{2}})T^n x\| \to 0$ as $\|(I - A)T^n x\| \to 0$.

Solution 6.2. Consider the properties (a), (b) and (c) in Problem 6.1.

(a) $A^{\frac{1}{2}} = O$ (equivalently, $A = O$) if and only if $T^n \xrightarrow{s} O$ by 6.1(b).

(b) Recall: T is an isometry if and only if $T^*T = I$, which turns out to be equivalent to $T^{*n}T^n = I$ for every $n \geq 1$. If $A = I$, then $T^*T = I$ by 6.1(c). Conversely, if $T^{*n}T^n = I$ for every $n \geq 1$, then $A = I$ by its very definition. That is, $A = I$ if and only if T is an isometry.

(c) Take an arbitrary vector x in $\mathcal{H}$. To begin with, consider the identity $\|ATx - TAx\|^2 = \|ATx\|^2 - 2\operatorname{Re}\langle ATx; TAx\rangle + \|TAx\|^2$. Now recall that $\langle ATx; TAx\rangle = \langle T^*ATx; Ax\rangle = \langle Ax; Ax\rangle = \|Ax\|^2$ according to 6.1(c). Moreover, since T is a contraction, $\|TAx\| \leq \|Ax\|$. Therefore,

$$\|ATx - TAx\|^2 \leq \|ATx\|^2 - \|Ax\|^2.$$

But T is a contraction and A is a nonnegative contraction (cf. 6.1(a)). Hence $\|T^*\| = \|T\| \leq 1$ and $\|A^{\frac{1}{2}}\| = \|A\|^{\frac{1}{2}} \leq 1$. Thus, by 6.1(c,d),

$$\|Ax\|^2 = \|T^*ATx\|^2 \leq \|ATx\|^2 = \|A^{\frac{1}{2}}A^{\frac{1}{2}}Tx\|^2 \leq \|A^{\frac{1}{2}}Tx\|^2 = \|A^{\frac{1}{2}}x\|^2.$$

If $A = A^2$, then $A^{\frac{1}{2}} = A$ (by uniqueness of the nonnegative square root) so that $\|Ax\|^2 = \|A^{\frac{1}{2}}x\|^2$. Thus the above displayed inequalities ensure that $\|Ax\|^2 = \|ATx\|^2$, and hence $\|ATx - TAx\|^2 = 0$. Therefore, $A = A^2$ implies $AT = TA$. On the other hand, if $AT = TA$, then $AT^n = T^nA$ for every $n \geq 1$. Thus, applying 6.1(c) once again, $A = T^{*n}AT^n = T^{*n}T^nA \xrightarrow{s} A^2$, and hence $A = A^2$. Summing up: $A = A^2$ if and only if $AT = TA$.

Solution 6.3. Take an arbitrary x in $\mathcal{H}$ and consider the properties (a), (b), (c), (d) and (g) in Problem 6.1. Recall that $\mathcal{N}(Q^{\frac{1}{2}}) = \mathcal{N}(Q)$ for every nonnegative operator Q.

(a) It is plain from 6.1(b) that

$$x \in \mathcal{N}(A^{\frac{1}{2}}) \quad \text{if and only if} \quad T^n x \to 0.$$

(b) Note that $(I - A)$ is a nonnegative operator by 6.1(a). Therefore, since $\|(I - A)^{\frac{1}{2}} x\|^2 = \langle (I - A)x \, ; x \rangle = \|x\|^2 - \|A^{\frac{1}{2}} x\|^2$, we get

$$x \in \mathcal{N}(I - A)^{\frac{1}{2}} \quad \text{if and only if} \quad \|A^{\frac{1}{2}} x\| = \|x\|.$$

Moreover, according to 6.1(d), $\|A^{\frac{1}{2}} x\| = \|A^{\frac{1}{2}} T^n x\| \le \|T^n x\| \le \|x\|$ for every $n \ge 1$ because $A^{\frac{1}{2}}$ and T are contractions. Thus, by 6.1(b),

$$\|A^{\frac{1}{2}} x\| = \|x\| \quad \text{if and only if} \quad \|T^n x\| = \|x\| \ \forall n \ge 1.$$

(c) The operator $(A - A^2)$ is nonnegative by 6.1(a) and Problem 2.3(b). Hence $\|(A - A^2)^{\frac{1}{2}} x\|^2 = \langle (A - A^2)x \, ; x \rangle = \|A^{\frac{1}{2}} x\|^2 - \|Ax\|^2$, and so

$$x \in \mathcal{N}(A - A^2)^{\frac{1}{2}} \quad \text{if and only if} \quad \|A^{\frac{1}{2}} x\| = \|Ax\|.$$

Moreover, 6.1(a), 6.1(c) and 6.1(d) ensure that

$$\|Ax\| = \|T^{*n} A T^n x\| \le \|A T^n x\| \le \|A^{\frac{1}{2}} T^n x\| = \|A^{\frac{1}{2}} x\|$$

for every $n \ge 1$. Thus, by 6.1(g),

$$\|A^{\frac{1}{2}} x\| = \|Ax\| \quad \text{if and only if} \quad \|A T^n x\| = \|Ax\| \ \forall n \ge 1.$$

Now $\|A T^n x - T^n A x\|^2 = \|A T^n x\|^2 - 2 \operatorname{Re} \langle T^{*n} A T^n x \, ; Ax \rangle + \|T^n Ax\|^2$ for each integer $n \ge 1$. Therefore, since T is a contraction, it follows by 6.1(c) that $\|A T^n x - T^n A x\|^2 \le \|A T^n x\|^2 - 2\|Ax\|^2 + \|Ax\|^2$. Then

$$\left(\|A T^n x\| - \|Ax\| \right)^2 \le \|A T^n x - T^n A x\|^2 \le \|A T^n x\|^2 - \|Ax\|^2,$$

for every $n \ge 1$, and hence

$$\|A T^n x\| = \|Ax\| \ \forall n \ge 1 \quad \text{if and only if} \quad A T^n x = T^n A x \ \forall n \ge 1.$$

Solution 6.4. Let A be any operator on a Hilbert space. Observe that the subspace $\mathcal{N}(A - A^2)$ is A-invariant. (Indeed, if $x \in \mathcal{N}(A - A^2)$, then $(A - A^2)x = 0$ so that $(I - A)Ax = 0$, and hence $Ax \in \mathcal{N}(I - A)$, which implies that Ax lies in $\mathcal{N}(A - A^2)$ because $\mathcal{N}(I - A) \subseteq \mathcal{N}(A - A^2)$.) Thus consider the restriction of A to the invariant subspace $\mathcal{N}(A - A^2)$,

$$P = A|_{\mathcal{N}(A - A^2)} : \mathcal{N}(A - A^2) \to \mathcal{N}(A - A^2).$$

Note that $P \in \mathcal{B}[\mathcal{N}(A - A^2)]$ is a projection ($P^2 x = A^2 x = Ax = Px$ for every x in $\mathcal{N}(A - A^2)$). Now suppose A is nonnegative. This implies that

P is an orthogonal projection (i.e., a nonnegative projection; see Problem 2.7). Indeed, P is nonnegative because A is ($\langle Px \,;x\rangle = \langle Ax \,;x\rangle$ for every x in $\mathcal{N}(A - A^2)$). Moreover,

$$\mathcal{N}(P) = \mathcal{N}(A|_{\mathcal{N}(A-A^2)}) = \mathcal{N}(A) \cap \mathcal{N}(A - A^2) = \mathcal{N}(A)$$

because $\mathcal{N}(A) \subseteq \mathcal{N}(A - A^2)$ and, since P is an orthogonal projection,

$$\begin{aligned}
\mathcal{N}(P)^\perp = \mathcal{R}(P) &= \{x \in \mathcal{N}(A - A^2)\colon Px = x\} \\
&= \{x \in \mathcal{N}(A - A^2)\colon Ax = x\} \\
&= \mathcal{N}(A - A^2) \cap \mathcal{N}(I - A) = \mathcal{N}(I - A)
\end{aligned}$$

because $\mathcal{N}(I - A) \subseteq \mathcal{N}(A - A^2)$. Therefore, by the Projection Theorem, the Hilbert space $\mathcal{N}(A - A^2)$ admits the orthogonal decomposition

$$\mathcal{N}(A - A^2) = \mathcal{N}(P) + \mathcal{N}(P)^\perp = \mathcal{N}(A) + \mathcal{N}(I - A).$$

Since $\mathcal{N}(A) \perp \mathcal{N}(I-A)$, which in fact happens for any self-adjoint operator, we may identify the orthogonal ordinary sum with the orthogonal direct sum (i.e., $\mathcal{N}(A) + \mathcal{N}(I - A) \cong \mathcal{N}(A) \oplus \mathcal{N}(I - A)$), and write

$$\mathcal{N}(A - A^2) = \mathcal{N}(A) \oplus \mathcal{N}(I - A).$$

Solution 6.5. If S is any linear transformation defined on an inner product space, then the linear transformation $S|_{\mathcal{N}(S)^\perp}$ is injective. Indeed, since $S|_{\mathcal{N}(S)^\perp}$ is linear, and since $\mathcal{N}(S|_{\mathcal{N}(S)^\perp}) \subseteq \mathcal{N}(S) \cap \mathcal{N}(S)^\perp = \{0\}$, it follows that $S|_{\mathcal{N}(S)^\perp}$ is injective (restriction of a linear transformation to a linear manifold is linear, and a linear transformation is injective if and only if its kernel is null). If, in addition, S is bounded and acts between Hilbert spaces, then $\mathcal{N}(S)^\perp = \mathcal{R}(S^*)^-$ (cf. Problem 2.1). In particular, if S is a self-adjoint operator on a Hilbert space $\mathcal{H}$, then $S|_{\mathcal{R}(S)^-}$ is injective.

Now take a contraction T in $\mathcal{B}[\mathcal{H}]$ and let A in $\mathcal{B}^+[\mathcal{H}]$ be the strong limit of $\{T^{*n}T^n\}$. Take the square root $A^{\frac{1}{2}}$ of A. Since $A^{\frac{1}{2}}$ is self-adjoint (it is nonnegative), it follows that $A^{\frac{1}{2}}|_{\mathcal{R}(A^{1/2})^-}$ is injective, and so is $A^{\frac{1}{2}}|_{\mathcal{R}(A^{1/2})}$. Consider the map $V_0 \colon \mathcal{R}(A^{1/2}) \to \mathcal{R}(A^{1/2})$ that assigns to each vector y in $\mathcal{R}(A^{1/2})$ the vector $V_0 y = A^{\frac{1}{2}} Tx$ in $\mathcal{R}(A^{1/2})$, where x is the *unique* vector in $\mathcal{H}$ such that $y = A^{\frac{1}{2}} x$. Since $A^{\frac{1}{2}} T$ is linear, it is readily verified that V_0 is linear. Moreover, note that

$$V_0 A^{\frac{1}{2}} x = A^{\frac{1}{2}} Tx,$$

and hence $\|V_0 A^{\frac{1}{2}} x\| = \|A^{\frac{1}{2}} Tx\| = \|A^{\frac{1}{2}} x\|$ (Problem 6.1(c)), for every x in $\mathcal{H}$. Then V_0 is an isometry on $\mathcal{R}(A^{1/2})$. Recall that $\mathcal{R}(A^{1/2})^- = \mathcal{R}(A)^-$ and extend V_0 by continuity to $\mathcal{R}(A^{1/2})^-$. Let the operator

$$V \colon \mathcal{R}(A)^- \to \mathcal{R}(A)^-$$

be such an extension, which is again an isometry. In fact, if x is any vector in $\mathcal{R}(A)^- = \mathcal{R}(A^{1/2})^-$, then $x = \lim x_n$ for some $\mathcal{R}(A^{1/2})$-valued sequence $\{x_n\}$, and so $\|Vx\| = \|V(\lim x_n)\| = \lim \|V_0 x_n\| = \lim \|x_n\| = \|x\|$. Thus for each contraction T in $\mathcal{B}[\mathcal{H}]$ there exists an isometry V in $\mathcal{B}[\mathcal{R}(A)^-]$ such that

$$A^{\frac{1}{2}}T = VA^{\frac{1}{2}}.$$

Solution 6.6. Consider the properties (a) and (c) of Problem 6.1. Suppose $A = A_*$. Thus $A = T^{*n}AT^n = T^{*n}A_*T^n = T^{*n}T^n A_* T^{*n}T^n$ for each $n \geq 1$. Hence $A = A^3$ (since $T^{*n}T^n \xrightarrow{s} A$; cf. Problem 3.2) so that $O \leq A^2 = A^4$. Therefore, by uniqueness of the nonnegative square root, $A = A^2$.

Solution 6.7. Let T be a contraction on a Hilbert space $\mathcal{H}$. Recall from Problem 6.3 that $\mathcal{N}(I - A) = \{x \in \mathcal{H}: \|T^n x\| = \|x\|$ for every $n \geq 1\}$ is T-invariant, and so $\mathcal{N}(I - A_*) = \{x \in \mathcal{H}: \|T^{*n}x\| = \|x\|$ for every $n \geq 1\}$ is T^*-invariant. Since intersection of subspaces is again a subspace,

$$\mathcal{U} = \mathcal{N}(I - A) \cap \mathcal{N}(I - A_*) = \{x \in \mathcal{H}: \|T^n x\| = \|T^{*n}x\| = \|x\| \ \forall n \geq 1\}$$

is an invariant subspace for both T and T^*. Hence (Problem 4.4 and 4.7) $\mathcal{U}$ reduces T so that $(T|_{\mathcal{U}})^* = T^*|_{\mathcal{U}}$. Then $\|(T|_{\mathcal{U}})x\| = \|(T|_{\mathcal{U}})^*x\| = \|x\|$ for every x in $\mathcal{U}$, which means that $T|_{\mathcal{U}}$ in $\mathcal{B}[\mathcal{U}]$ is unitary (Problem 2.6). If a subspace $\mathcal{M}$ reduces T, then $(T|_{\mathcal{M}})^n x = T^n x$ and $(T|_{\mathcal{M}})^{*n}x = T^{*n}x$ for every x in $\mathcal{M}$ and every $n \geq 1$. If, in addition, $T|_{\mathcal{M}}$ in $\mathcal{B}[\mathcal{M}]$ is unitary, then $\|(T|_{\mathcal{M}})^n x\| = \|(T|_{\mathcal{M}})^{*n}x\| = \|x\|$ so that $\|T^n x\| = \|T^{*n}x\| = \|x\|$ for every x in $\mathcal{M}$ and every $n \geq 1$. Thus $\mathcal{M} \subseteq \mathcal{U}$. Therefore, $\mathcal{U}$ is the largest reducing subspace for T on which T is unitary, and hence $T|_{\mathcal{U}^\perp}$ must be completely nonunitary; completing the proof of the Nagy–Foiaş–Langer decomposition.

Solution 6.8. If T is an isometry (i.e., if $A = I$), then it is a contraction with $\mathcal{N}(I - A) = \mathcal{H}$, and hence $\mathcal{N}(I - A) \cap \mathcal{N}(I - A_*) = \mathcal{N}(I - A_*)$. Apply the Nagy–Foiaş–Langer decomposition with $\mathcal{U} = \mathcal{N}(I - A_*)$ so that $T|_{\mathcal{U}^\perp}$ is a completely nonunitary isometry on $\mathcal{U}^\perp$, which means a unilateral shift.

Solution 6.9. Let T be a contraction on a Hilbert space $\mathcal{H}$. As usual, identify ordinary sum of orthogonal subspaces with orthogonal direct sum.

(a) If $A = A^2$, then $\mathcal{H} = \mathcal{N}(A - A^2) = \mathcal{N}(A) \oplus \mathcal{N}(I - A)$, where $\mathcal{N}(A)$ and $\mathcal{N}(I - A)$ are orthogonal complementary T-invariant subspaces of $\mathcal{H}$ (see Problems 6.3 and 6.4), and hence they reduce T. Thus

$$T = G \oplus J,$$

where (Problem 6.3) $G = T|_{\mathcal{N}(A)}$ is a strongly stable contraction on $\mathcal{N}(A)$ and $J = T|_{\mathcal{N}(I-A)}$ is an isometry on $\mathcal{N}(I-A)$. But the von Neumann–Wold decomposition says that $J = S_+ \oplus U$, and so

$$T = G \oplus S_+ \oplus U,$$

where $S_+ = J|_{\mathcal{M}}$ is a unilateral shift on $\mathcal{M}$ and $U = J|_{\mathcal{U}}$ is unitary on $\mathcal{U}$, $\mathcal{M}$ and $\mathcal{U}$ being orthogonal complementary subspaces of $\mathcal{N}(I - A)$; that is, $\mathcal{N}(I - A) = \mathcal{M} \oplus \mathcal{U}$. Since $G \oplus S_+$ is completely nonunitary (a strongly stable contraction and a unilateral shift are completely nonunitary, and so their direct sum is a completely nonunitary contraction), and since the Nagy–Foiaş–Langer decomposition for T is unique, it follows that

$$\mathcal{U} = \mathcal{N}(I - A) \cap \mathcal{N}(I - A_*).$$

Claim. $\qquad\qquad\qquad \mathcal{M} = \mathcal{N}(I - A) \cap \mathcal{N}(A_*).$

Proof. Note that $T^* = G^* \oplus S_+^* \oplus U^*$ on $\mathcal{H} = \mathcal{N}(A) \oplus \mathcal{M} \oplus \mathcal{U}$, and recall from Problem 5.1 that $S_+^{*n} \xrightarrow{s} O$. Therefore, $\mathcal{M} \subseteq \mathcal{N}(A_*)$ according to Problem 6.3. But $\mathcal{M} = \mathcal{N}(I - A) \ominus \mathcal{U}$ so that $\mathcal{M} \subseteq \mathcal{N}(I - A)$. Hence

$$\mathcal{M} \subseteq \mathcal{N}(I - A) \cap \mathcal{N}(A_*).$$

On the other hand, since U^* is unitary, $\mathcal{N}(A_*) \subseteq \mathcal{N}(A) \oplus \mathcal{M}$ (cf. Problem 6.3 again), and so

$$\mathcal{N}(I - A) \cap \mathcal{N}(A_*) \subseteq \mathcal{N}(I - A) \cap (\mathcal{N}(A) \oplus \mathcal{M}) \subseteq \mathcal{M}$$

because $\mathcal{N}(A) \perp \mathcal{N}(I - A)$ and $\mathcal{M} \subseteq \mathcal{N}(I - A)$. Indeed, if x is an arbitrary vector in $\mathcal{N}(I - A) \cap (\mathcal{N}(A) \oplus \mathcal{M})$, then $x \in \mathcal{N}(I - A)$ and $x = (u, v)$, with u in $\mathcal{N}(A)$ and v in $\mathcal{M}$. Since $\mathcal{N}(A) \perp \mathcal{N}(I - A)$ and since $\mathcal{M} \subseteq \mathcal{N}(I - A)$, it follows that $\mathcal{N}(A) \perp \mathcal{M}$. Thus we may identify u in $\mathcal{N}(A)$ and v in $\mathcal{M}$ with $(u, 0)$ and $(0, v)$, respectively, in $\mathcal{N}(A) \oplus \mathcal{M}$. Therefore, $0 = \langle u ; x \rangle = \langle (u, 0) ; (u, v) \rangle = \langle u ; u \rangle + \langle 0 ; v \rangle = \|u\|^2$, and hence $x = (0, v)$ lies in $\mathcal{M}$. $\square$

Conversely, note that $T = G \oplus S_+ \oplus U$ implies $A = O \oplus I \oplus I$ (see Problem 6.2), which in turn implies $A = A^2$ and completes the proof of (a).

(b) Since G is a contraction on $\mathcal{N}(A)$, let the operator A'_* on $\mathcal{N}(A)$ be the strong limit of $\{G^n G^{*n}\}$. But $T^n T^{*n} = G^n G^{*n} \oplus S_+^n S_+^{*n} \oplus U^n U^{*n}$ for each integer $n \geq 0$, and so $A_* = A'_* \oplus O \oplus I$, on $\mathcal{H} = \mathcal{N}(A) \oplus \mathcal{M} \oplus \mathcal{U}$ (reason: S_+^* is a strongly stable contraction on $\mathcal{M}$ and U^* is an isometry on $\mathcal{U}$). Therefore, $\mathcal{N}(A'_*) \subseteq \mathcal{N}(A)$ and $\mathcal{N}(A_*) = \mathcal{N}(A'_*) \oplus \mathcal{M}$, and hence

$$\mathcal{N}(A) \cap \mathcal{N}(A_*) = \mathcal{N}(A) \cap (\mathcal{N}(A'_*) \oplus \mathcal{M}) = \mathcal{N}(A) \cap \mathcal{N}(A'_*) = \mathcal{N}(A'_*)$$

because $\mathcal{N}(A) \perp \mathcal{M}$. Similarly, $(I - A_*) = (I - A'_*) \oplus I \oplus O$, which implies that $\mathcal{N}(I - A'_*) \subseteq \mathcal{N}(A)$ and $\mathcal{N}(I - A_*) = \mathcal{N}(I - A'_*) \oplus \mathcal{U}$. Thus

$$\mathcal{N}(A) \cap \mathcal{N}(I - A_*) = \mathcal{N}(A) \cap (\mathcal{N}(I - A'_*) \oplus \mathcal{U}) = \mathcal{N}(I - A'_*)$$

because $\mathcal{N}(A) \perp \mathcal{U}$ (recall: $\mathcal{U} = \mathcal{N}(I - A) \ominus \mathcal{M}$ and $\mathcal{N}(A) \perp \mathcal{N}(I - A)$). Now suppose, in addition to $A = A^2$, that $A_* = A_*^2$. Then $A'_* = A'^2_*$, and hence $\mathcal{N}(A) = \mathcal{N}(A'_* - A'^2_*) = \mathcal{N}(A'_*) \oplus \mathcal{N}(I - A'_*)$, where $\mathcal{N}(A'_*)$ and $\mathcal{N}(I - A'_*)$ are orthogonal complementary G^*-invariant subspaces of $\mathcal{N}(A)$ (see Problems 6.3 and 6.4). Thus they reduce G^* and so they reduce G. Then $G = B \oplus S_-$ so that

$$T = B \oplus S_- \oplus S_+ \oplus U,$$

where (cf. Problem 6.3) $B = G|_{\mathcal{N}(A'_*)} = T|_{\mathcal{N}(A) \cap \mathcal{N}(A_*)}$ is a strongly stable contraction on $\mathcal{N}(A) \cap \mathcal{N}(A_*)$ whose adjoint B^* also is strongly stable, and the other direct summand $S_- = G|_{\mathcal{N}(I - A'_*)} = T|_{\mathcal{N}(A) \cap \mathcal{N}(I - A_*)}$ is a strongly stable (so completely nonunitary) contraction on $\mathcal{N}(A) \cap \mathcal{N}(I - A_*)$ whose adjoint is a (completely nonunitary) isometry. Therefore, S_-^* is a unilateral shift, which means that its adjoint S_- is a backward unilateral shift. On the other hand, if $T = B \oplus S_- \oplus S_+ \oplus U$, then it follows from Problem 6.2 that $A = O \oplus O \oplus I \oplus I$ and $A_* = O \oplus I \oplus O \oplus I$, which implies $A = A^2$ and $A_* = A_*^2$, thus completing the proof of (b).

(c) If $A = A_*$, then $A = A^2$ and $A_* = A_*^2$ (see Problem 6.6) so that T can be decomposed as in (b). Moreover, since $\mathcal{N}(A) \perp \mathcal{N}(I - A)$ (because A is self-adjoint), it follows that $\mathcal{N}(I - A) \cap \mathcal{N}(A_*) = \mathcal{N}(A) \cap \mathcal{N}(I - A_*) = \{0\}$, and hence the decomposition in (b) is reduced to

$$T = B \oplus U,$$

on $\mathcal{H} = \mathcal{N}(A) \oplus \mathcal{N}(I - A)$, with $B = T|_{\mathcal{N}(A)}$ and $U = T|_{\mathcal{N}(I-A)}$, where B is a strongly stable contraction on $\mathcal{N}(A)$ such that B^* also is strongly stable, and U is a unitary operator on $\mathcal{N}(I - A)$. Conversely, if $T = B \oplus U$, then $A = A_* = O \oplus I$ (cf. Problem 6.1), which completes the proof of (c).

Solution 6.10. The Foguel decomposition says that *every contraction is either a weakly stable contraction, a unitary operator, or a direct sum of a weakly stable contraction and a unitary operator.* Thus a completely nonunitary contraction (i.e., a contraction that has no nonzero unitary direct summand) must be weakly stable.

Solution 6.11. Let T in $\mathcal{B}[\mathcal{H}]$ be an isometry on a Hilbert space $\mathcal{H}$ and let J in $\mathcal{B}[\mathcal{K}]$ be an isometry on a Hilbert space $\mathcal{K}$. Thus $T^*T = I$ (identity on $\mathcal{H}$) and $J^*J = I$ (identity on $\mathcal{K}$). If there exists X in $\mathcal{B}[\mathcal{H}, \mathcal{K}]$ such that $XT^* = JX$, then $XT = J^*JXT = J^*XT^*T = J^*X$, and therefore T has property PF. That is, every isometry has property PF.

Solution 6.12. Suppose a contraction T on a Hilbert space $\mathcal{H}$ has a co-isometry as a direct summand. That is, suppose there exists a proper subspace $\mathcal{M}$ of $\mathcal{H}$ that reduces T for which $T = S \oplus J^*$, where S is an operator on $\mathcal{M}$ and J is an isometry on $\mathcal{M}^\perp$. Set $X = O \oplus J$ and $W = I \oplus J$ on $\mathcal{H} = \mathcal{M} \oplus \mathcal{M}^\perp$. Since

$$XT^* = O \oplus J^2 = WX,$$

X intertwines T^* to the isometry W. If T has property PF, then

$$O \oplus JJ^* = XT = W^*X = O \oplus J^*J$$

so that J is a normal isometry (i.e., a unitary operator). Summing up: if an isometry J is such that J^* is a direct summand of a contraction with property PF, then J is unitary.

Solution 6.13. Take an arbitrary isometry J in $\mathcal{B}[\mathcal{K}]$ so that $\|J^n y\| = \|y\|$ for every positive integer n and every y in $\mathcal{K}$. If $T^n \xrightarrow{s} O$ and $XT = JX$ for some X in $\mathcal{B}[\mathcal{H}, \mathcal{K}]$, then $\|Xx\| = \|J^n Xx\| = \|XT^n x\| \to 0$ for every x in $\mathcal{H}$, and hence $X = O$. Conversely, recall that $A^{\frac{1}{2}} T = VA^{\frac{1}{2}}$ (cf. Problem 6.5). If $T^n \xrightarrow{s} \!\!\!\!\!/\; O$ (i.e., if $A \neq O$), then set $\mathcal{K} = \mathcal{R}(A^{\frac{1}{2}})^- = \mathcal{R}(A)^- \neq \{0\}$ and consider the transformation X in $\mathcal{B}[\mathcal{H}, \mathcal{K}]$ defined by $Xx = A^{\frac{1}{2}} x$ for every vector x in $\mathcal{H}$. Thus $XT = VX$ and $X \neq O$. Therefore, if $XT = JX$ implies $X = O$ for every isometry J on any Hilbert space $\mathcal{K}$, then $T^n \xrightarrow{s} O$.

Solution 6.14. Let T be a contraction on a Hilbert space $\mathcal{H}$ so that T^* is a contraction as well. Consider the nonnegative contraction A_* on $\mathcal{H}$ and the isometry V_* on $\mathcal{R}(A_*)^-$ such that $A_*^{1/2} T^* = V_* A_*^{1/2}$ (cf. Problem 6.5). Take an arbitrary nonnegative integer n. Thus

$$A_*^{1/2} T^{*n} = V_*^n A_*^{1/2}.$$

If T has property PF, then $A_*^{1/2} T = V_*^* A_*^{1/2}$ so that $A_*^{1/2} T^n = V_*^{*n} A_*^{1/2}$, and hence

$$A_*^{1/2} V_*^n = T^{*n} A_*^{1/2}$$

because $A_*^{1/2}$ is self-adjoint. But $A_* = T^n A_* T^{*n}$ by Problem 6.1(c) so that

$$A_* = T^n A_*^{1/2} A_*^{1/2} T^{*n} = T^n A_*^{1/2} V_*^n A_*^{1/2} = T^n T^{*n} A_*^{1/2} A_*^{1/2},$$

and therefore $A_* = A_*^2$ (reason: $T^n T^{*n} \xrightarrow{s} A_*$). Note: The above solution was borrowed from [35]; a different one can be found in [58].

Solution 6.15. If a contraction T on a Hilbert space $\mathcal{H}$ is such that T^* has property PF, then A is a projection by Problem 6.14. Hence T is a strongly stable contraction G, or a unilateral shift S_+, or a unitary operator U, or a direct sum of any of them (Problem 6.9). Thus

$$T^* = G^* \oplus S_+^* \oplus U^*,$$

Since T has property PF, Problem 6.12 ensures that S_+^* cannot be present in the above decomposition. Therefore,

$$T^* = G^* \oplus U^*,$$

and the completely nonunitary direct summand G of $T = G \oplus U$ is strongly stable (recall that G is completely nonunitary because it is strongly stable, and any direct summand of U is again unitary). Conversely, consider the Nagy–Foiaş–Langer decomposition for the contraction T (Problem 6.7),

$$T = U \oplus C$$

on $\mathcal{H} = \mathcal{U} \oplus \mathcal{U}^\perp$, where $U = T|_\mathcal{U}$ in $\mathcal{B}[\mathcal{U}]$ is unitary and $C = T|_{\mathcal{U}^\perp}$ in $\mathcal{B}[\mathcal{U}^\perp]$ is a completely nonunitary contraction (the completely nonunitary direct summand of T). Let $\mathcal{K}$ be a Hilbert space and suppose $XT = JX$ for some isometry J in $\mathcal{B}[\mathcal{K}]$ and some X in $\mathcal{B}[\mathcal{H}, \mathcal{K}]$. The von Neumann–Wold decomposition for isometries (Problem 6.8) says that

$$J = W \oplus S_+$$

on $\mathcal{K} = \mathcal{W} \oplus \mathcal{W}^\perp$, where $W = J|_\mathcal{W}$ in $\mathcal{B}[\mathcal{W}]$ is unitary, and $S_+ = J|_{\mathcal{W}^\perp}$ in $\mathcal{B}[\mathcal{W}^\perp]$ is a unilateral shift. Now write X with respect to the decompositions $\mathcal{H} = \mathcal{U} \oplus \mathcal{U}^\perp$ and $\mathcal{K} = \mathcal{W} \oplus \mathcal{W}^\perp$,

$$X = \begin{pmatrix} X_{11} & X_{12} \\ X_{21} & X_{22} \end{pmatrix},$$

with X_{11} in $\mathcal{B}[\mathcal{U}, \mathcal{W}]$, X_{12} in $\mathcal{B}[\mathcal{U}^\perp, \mathcal{W}]$, X_{21} in $\mathcal{B}[\mathcal{U}, \mathcal{W}^\perp]$ and X_{22} in $\mathcal{B}[\mathcal{U}^\perp, \mathcal{W}^\perp]$. Since $XT = JX$ we get

$$\begin{array}{ll} X_{11}U = WX_{11}, & X_{12}C = WX_{12}, \\ X_{21}U = S_+X_{21}, & X_{22}C = S_+X_{22}. \end{array}$$

Problem 6.11 ensures that

$$X_{11}U^* = W^*X_{11}$$

because U^* and W are isometries. Since $X_{21}^*S_+^* = U^*X_{21}^*$ and S_+^* is strongly stable, it follows by Problem 6.13 that $X_{21}^* = O$, and hence $X_{21} = O$. Suppose C is strongly stable. Since W and S_+ are isometries, it also follows by Problem 6.13 that $X_{12} = O$ and $X_{22} = O$. Thus $XT^* = J^*X$, and so T^* has property PF.

7

Hyponormal Operators

Take any operator T on a Hilbert space $\mathcal{H}$. Set

$$D_T = [T^*, T] = T^*T - TT^*$$

in $\mathcal{B}[\mathcal{H}]$, which is always self-adjoint. This is called the *self-commutator* of T. Recall that T is normal if it commutes with its adjoint. In other words,

$$T \text{ is normal if and only if } D_T = O.$$

An operator $T \in \mathcal{B}[\mathcal{H}]$ is *quasinormal* if it commutes with T^*T. Equivalently, if $(T^*T - TT^*)T = O$. Therefore,

$$T \text{ is quasinormal if and only if } D_T T = O.$$

It is obvious that *every normal operator is quasinormal*. Observe that *every isometry is quasinormal*. (Indeed, if V is an isometry on a Hilbert space, then $V^*V = I$ so that $V^*VV - VV^*V = O$.)

Problem 7.1. Show that a direct summand of a quasinormal operator is again quasinormal (the restriction of a quasinormal operator to a reducing subspace is quasinormal). If T is quasinormal, then $\mathcal{N}(T)$ reduces T, and hence $T|_{\mathcal{N}(T)^\perp}$ is quasinormal.

Problem 7.2. Take an arbitrary nonnegative integer n. Show that if T is a quasinormal operator, then

(a) $(T^*T)^nT = T(T^*T)^n$,

(b) $T^{*n}T^n = (T^*T)^n$,

(c) $(T^{*n}T^n)T = T(T^{*n}T^n)$,

(d) $|T^n| = |T|^n$.

Moreover, also show that

(e) $T^n \xrightarrow{s} O \iff |T|^n \xrightarrow{s} O \iff |T|^n \xrightarrow{w} O$.

For a survey on quasinormal operators see [54]. A *part* of an operator is a restriction of it to an invariant subspace. An operator is *subnormal* if it is a part of a normal operator; that is, if it has a normal extension. In other words, an operator T on a Hilbert space $\mathcal{H}$ is subnormal if there exists a Hilbert space $\mathcal{K}$ including $\mathcal{H}$ and a normal operator N on $\mathcal{K}$ such that $\mathcal{H}$ is N-invariant ($N(\mathcal{H}) \subseteq \mathcal{H}$) and T is the restriction of N to $\mathcal{H}$ ($T = N|_{\mathcal{H}}$). Thus T in $\mathcal{B}[\mathcal{H}]$ is subnormal if $\mathcal{H}$ is a subspace of $\mathcal{K}$ and, with respect to the decomposition $\mathcal{K} = \mathcal{H} \oplus \mathcal{H}^{\perp}$ (cf. Problem 4.2),

$$N = \begin{pmatrix} T & X \\ O & Y \end{pmatrix}$$

in $\mathcal{B}[\mathcal{K}]$ is normal for some $X \in \mathcal{B}[\mathcal{H}^{\perp}, \mathcal{H}]$ and $Y \in \mathcal{B}[\mathcal{H}^{\perp}]$. For a treatise on subnormal operators the reader is referred to [9]. It is not difficult to verify that *every quasinormal operator is subnormal* (see e.g., [9, p. 29] or [32, p. 445]). An operator T is *hyponormal* if $TT^* \leq T^*T$. That is,

$$T \text{ is hyponormal if and only if } D_T \geq O.$$

Problem 7.3. Show that an operator T on a Hilbert space $\mathcal{H}$ is hyponormal if and only if $\|T^*x\| \leq \|Tx\|$ for every $x \in \mathcal{H}$.

An operator $T \in \mathcal{B}[\mathcal{H}]$ is *cohyponormal* if its adjoint $T^* \in \mathcal{B}[\mathcal{H}]$ is hyponormal (i.e., if $T^*T \leq TT^*$ or, equivalently, if $D_T \leq O$, which means by the above problem that $\|Tx\| \leq \|T^*x\|$ for every $x \in \mathcal{H}$). Hence T *is normal if and only if it is both hyponormal and cohyponormal* (cf. Problems 2.4 and 7.3). If an operator is either hyponormal or cohyponormal, then it is called *seminormal*. It is easy to show that *every subnormal operator is hyponormal* (see e.g., [9, p. 46] or [32, p. 446]). Therefore, every quasinormal operator is hyponormal. In the next problem we ask for a direct proof, without passing through subnormal operators.

Problem 7.4. A quasinormal operator is hyponormal. Give a direct proof.

Problem 7.5. Let $\mathcal{H}$ be a Hilbert space. Take any operator T in $\mathcal{B}[\mathcal{H}]$ and any vector x in $\mathcal{H}$. Show that

(a) $T^*Tx = \|T\|^2 x$ if and only if $\|Tx\| = \|T\| \|x\|$.

Now consider the subset

$$\mathcal{M} = \{x \in \mathcal{H} \colon \|Tx\| = \|T\| \|x\|\}$$

of $\mathcal{H}$ and prove the following assertions.

(b) $\mathcal{M}$ is a subspace of $\mathcal{H}$.

(c) If T is hyponormal, then $\mathcal{M}$ is T-invariant.

(d) If T is normal, then $\mathcal{M}$ reduces T.

Note that $\mathcal{M}$ may be trivial — samples: $T = I$ and $T = \mathrm{diag}\big(\big\{\frac{k+1}{k+2}\big\}_{k=0}^{\infty}\big)$.

Problem 7.6. Let $\mathcal{M}$ be an invariant subspace for an operator T in $\mathcal{B}[\mathcal{H}]$.

(a) If T is hyponormal, then $T|_{\mathcal{M}}$ is hyponormal.

(b) If T is hyponormal and $T|_{\mathcal{M}}$ is normal, then $\mathcal{M}$ reduces T.

In the previous problem we saw that the restriction of a hyponormal to an invariant subspace is again hyponormal and, if this restriction is normal, then the invariant subspace actually is a reducing one. The next problem says that the restriction of a normal operator to an invariant subspace is normal if and only if the invariant subspace is a reducing one.

Problem 7.7. Let $\mathcal{M}$ be an invariant subspace for a normal operator T in $\mathcal{B}[\mathcal{H}]$. Show that $T|_{\mathcal{M}}$ is normal if and only if $\mathcal{M}$ reduces T.

Problem 7.8. If $\mathcal{X}$ is a normed space and T is an operator in $\mathcal{B}[\mathcal{X}]$, then show that the real-valued sequence $\{\|T^n\|^{\frac{1}{n}}\}$ converges in $\mathbb{R}$.

Notation: Let the limit of $\{\|T^n\|^{\frac{1}{n}}\}$ be denoted by $r(T)$;

$$r(T) = \lim_n \|T^n\|^{\frac{1}{n}}.$$

Remarks: Since $\limsup_n \|T^n\|^{\frac{1}{n}} \leq \|T^m\|^{\frac{1}{m}}$ for every $m \geq 1$ (Solution 7.8), we get $r(T) \leq \|T^n\|^{\frac{1}{n}}$ for every $n \geq 1$ and, in particular, $r(T) \leq \|T\|$. Note that $r(T^k)^{\frac{1}{k}} = \lim_n \|T^{kn}\|^{\frac{1}{kn}} = r(T)$ for each $k \geq 1$ because $\{\|T^{kn}\|^{\frac{1}{kn}}\}$ is a subsequence of the convergent sequence $\{\|T^n\|^{\frac{1}{n}}\}$. Thus $r(T^k) = r(T)^k$ for each $k \geq 1$. Therefore, if T is any operator on a normed space $\mathcal{X}$ and n is an arbitrary nonnegative integer, then

$$r(T)^n = r(T^n) \leq \|T^n\| \leq \|T\|^n.$$

Definition: If $r(T) = \|T\|$, then $T \in \mathcal{B}[\mathcal{X}]$ is a *normaloid* operator.

Problem 7.9. An operator T on a normed space $\mathcal{X}$ is normaloid if and only if $\|T^n\| = \|T\|^n$ for every integer $n \geq 0$.

Problem 7.10. Let T be an operator on Hilbert space. Prove the following propositions.

(a) If T is hyponormal, then $\|T^n\|^2 \leq \|T^{n+1}\| \|T^{n-1}\|$ for every $n \geq 1$.

(b) If $\|T^n\|^2 \leq \|T^{n+1}\| \|T^{n-1}\|$ for every $n \geq 1$, then $\|T^n\| = \|T\|^n$ for every $n \geq 0$.

(c) Every hyponormal operator is normaloid.

Since $\|T^{*n}\| = \|T^n\|$ for each $n \geq 0$, it follows that $r(T^*) = r(T)$. Thus T is normaloid if and only if T^* is normaloid, and hence every seminormal operator is normaloid. Summing up: An operator T is normal if it commutes with its adjoint, quasinormal if it commutes with T^*T, subnormal if it is a restriction of a normal operator to an invariant subspace, hyponormal if $TT^* \leq T^*T$, and normaloid if $r(T) = \|T\|$. These classes are related by proper inclusion as follows.

$$\text{Normal} \subset \text{Quasinormal} \subset \text{Subnormal} \subset \text{Hyponormal} \subset \text{Normaloid.}$$

Example: The above inclusions are, in fact, proper. The unilateral shift will do the whole job. First recall that a unilateral shift S_+ is an isometry but not a coisometry, and so S_+ is a nonnormal quasinormal operator. Since S_+ is subnormal, $A = I + S_+$ is subnormal (if N is a normal extension of S_+, then $I + N$ is a normal extension of A). However, since S_+ is a nonnormal isometry, $A^*AA - AA^*A = A^*AS_+ - S_+A^*A = S_+^*S_+ - S_+S_+^* \neq O$, and therefore A is not quasinormal. Put $B = S_+^* + 2S_+$. It can be shown that B is hyponormal but B^2 is not hyponormal [22, p. 313], and it is readily verified that the square of every subnormal operator is again subnormal. Thus the hyponormal B is not subnormal. Finally, S_+^* is normaloid (since S_+ is) but not hyponormal (because S_+ is a nonnormal hyponormal).

Remark: This chapter tackled only a few elementary aspects of hyponormal operators that will be required in the sequel. Further properties will be considered in each of the subsequent chapters. A comprehensive account on the theory of hyponormal operators can be found in [7], [9], [41] and [55] (see also [3], [18] and [32]).

We close the chapter with unilateral weighted shifts. It is very easy to characterize quasinormal and hyponormal unilateral weighted shifts, as we shall see in Chapter 9. However, for subnormal weighted shifts the task is not so easy. Let us first verify unitary equivalence for subnormal operators.

Problem 7.11. Let T be an operator on a Hilbert space. Show that if T is subnormal, then so is every operator unitarily equivalent to it.

Now let T_+ be a unilateral weighted shift with a nonnegative weight sequence $\{\alpha_k\}_{k=0}^{\infty}$ (no loss of generality according to Problems 5.10 and 7.11). Suppose T_+ is injective so that its weight sequence is, in fact, positive (Problem 5.9). Again, without loss of generality, assume that $\|T_+\| = 1$. It can be shown (see e.g., [9, p. 57], [21, p. 895] or [50, p. 84]) that T_+ *is subnormal if and only if there exists a probability measure* ν *on* $[0,1]$ *containing* 1 *in its support such that*

$$\prod_{k=0}^{n-1} \alpha_k^2 = \int x^{2n} d\nu(x) \quad \text{for every} \quad n \geq 1.$$

Remark: The support $[\nu]$ of a probability measure ν on $[0,1]$ is the smallest (in the inclusion ordering) closed Borel subset of $[0,1]$ such that $\nu([\nu]) = 1$. A probability measure ν on $[0,1]$ contains 1 in its support if and only if $\nu([1-\varepsilon, 1]) > 0$ for every $\varepsilon > 0$.

Problem 7.12. Let $T_+ = \text{shift}(\{\alpha_k\}_{k=0}^{\infty})$ be a unilateral weighted shift on ℓ_+^2 with nonnegative weights. Suppose T_+ is injective and $\|T_+\| = 1$ (so that $0 < \alpha_k \leq 1 = \sup_{k \geq 0} |\alpha_k|$ for every $k \geq 0$). Use the above result to show that, if T_+ is subnormal and $\alpha_m = \alpha_{m+1}$ for some $m \geq 1$, then $\alpha_k = 1$ for all $k \geq 1$. That is,
$$T_+ = \text{shift}(\alpha_0, 1, 1, 1, \ldots).$$

Conversely, show that $T_+ = \text{shift}(\alpha, 1, 1, 1, \ldots)$ is subnormal for all $\alpha \in (0,1]$.

Problem 7.13. Let $T_+ = \text{shift}(\{\alpha_k\}_{k=0}^{\infty})$ be a unilateral weighted shift on ℓ_+^2 such that $0 < \alpha_k \leq 1 = \sup_{k \geq 0} |\alpha_k|$ for every $k \geq 0$. If $\alpha_0 = \alpha_1 \neq 1$, then show that T_+ is not subnormal.

In Problems 7.12 and 7.13 we met weight sequences with at least one repeated weight. The next problem exhibits a subnormal unilateral weighted shift with distinct weights.

Problem 7.14. The following unilateral weighted shift on ℓ_+^2 is subnormal:
$$T_+ = \text{shift}\left(\left\{\left(\tfrac{k+1}{k+2}\right)^{\frac{1}{2}}\right\}_{k=0}^{\infty}\right).$$

Solutions

Solution 7.1. Let $\mathcal{M}$ be a reducing subspace for an operator T acting on a Hilbert space $\mathcal{H}$ so that $T = A \oplus D$ on $\mathcal{H} = \mathcal{M} \oplus \mathcal{M}^{\perp}$, where $A = T|_{\mathcal{M}}$ and $D = T|_{\mathcal{M}^{\perp}}$. Suppose T is quasinormal. Then

$$T^*TT = A^*AA \oplus D^*DD = AA^*A \oplus DD^*D = TT^*T$$

so that $A^*AA = AA^*A$ and $D^*DD = DD^*D$, and hence A and D are both quasinormal. Moreover, since T^*T is self-adjoint and commutes with T, then it also commutes with T^*. Therefore, $\mathcal{N}(T^*T)$ reduces T (cf. Problem 4.5). But we know that $\mathcal{N}(T^*T) = \mathcal{N}(T)$ by Problem 2.1 so that $\mathcal{N}(T)$ reduces T, and so the direct summand $D = T|_{\mathcal{N}(T)^\perp}$ is quasinormal.

Solution 7.2. Let T be a quasinormal operator so that $(T^*T)T = T(T^*T)$. Consider the assertions (a)–(e) in the problem statement.

(a) Assertion (a) holds trivially for $n = 0$ for any operator T. Suppose T is quasinormal, which means that (a) holds for $n = 1$. If (a) holds for some integer $n \geq 1$, then (since it holds for $n = 1$),

$$(T^*T)^{n+1}T = (T^*T)(T^*T)^nT = (T^*T)T(T^*T)^n = T(T^*T)^{n+1},$$

and hence (a) holds for $n+1$, which completes the proof by induction.

(b) Assertion (b) holds trivially for $n = 0, 1$ for any operator T. Let T be a quasinormal operator (so that assertion (a) holds true). Suppose $T^{*n}T^n = (T^*T)^n$ for some $n \geq 1$. Then (since (a) holds)

$$T^{*n+1}T^{n+1} = T^*T^{*n}T^nT = T^*(T^*T)^nT = T^*T(T^*T)^n = (T^*T)^{n+1},$$

which completes the proof by induction.

(c) This is straightforward by assertions (a) and (b).

(d) Take an arbitrary $n \geq 0$. Assertion (b) says that $|T^n|^2 = |T|^{2n}$. Then $|T^n| = |T|^n$ by uniqueness of the nonnegative square root once any positive integral power of a nonnegative operator is again nonnegative (reason: If $Q \geq O$ and $R \geq O$ commute, then $QR = R^{\frac{1}{2}}QR^{\frac{1}{2}} \geq O$).

(e) This is straightforward by Problem 3.9(d) and assertion (d).

Solution 7.3. Take $T \in \mathcal{B}[\mathcal{H}]$, $\mathcal{H}$ is a Hilbert space. $TT^* \leq T^*T$ if and only if $\langle TT^*x \, ; x \rangle \leq \langle T^*Tx \, ; x \rangle$; that is, $\|T^*x\|^2 \leq \|Tx\|^2$, for every $x \in \mathcal{H}$.

Solution 7.4. Let T be an operator on a Hilbert space $\mathcal{H}$. Recall that $\mathcal{N}(T^*)^\perp$ is a subspace of $\mathcal{H}$ that coincides with $\mathcal{R}(T)^-$ (see Problem 2.1), and therefore $\mathcal{H} = \mathcal{N}(T^*) + \mathcal{N}(T^*)^\perp = \mathcal{N}(T^*) + \mathcal{R}(T)^-$ by the Projection Theorem. Take an arbitrary x in $\mathcal{H}$ and write $x = u + v$ with u in $\mathcal{N}(T^*)$ and v in $\mathcal{R}(T)^-$. Consider the self-commutator $D_T = T^*T - TT^*$ on $\mathcal{H}$.

$$\langle D_T u \, ; u \rangle = \|Tu\|^2 - \|T^*u\|^2 = \|Tu\|^2$$

because u lies in $\mathcal{N}(T^*)$. Since v lies in the closure of $\mathcal{R}(T)$, it is the limit of an $\mathcal{R}(T)$-valued sequence, say $\{v_n\}$. Now suppose T is quasinormal so that $D_T T = O$. Then $D_T(\mathcal{R}(T)) = \{0\}$, and hence $D_T v_n = 0$ for all n. As D_T is continuous and self-adjoint, and by continuity of the inner product,

$$\langle D_T v \, ; v \rangle = \lim_n \langle D_T v_n \, ; v \rangle = 0,$$

$$\langle D_T u \, ; v \rangle = \lim_n \langle u \, ; D_T v_n \rangle = 0 \quad \text{and} \quad \langle D_T v \, ; u \rangle = \lim_n \langle D_T v_n \, ; u \rangle = 0.$$

Thus $\langle D_T x \, ; x \rangle = \langle D_T u \, ; u \rangle + \langle D_T u \, ; v \rangle + \langle D_T v \, ; u \rangle + \langle D_T v \, ; v \rangle = \|Tu\|^2 \geq 0$, and therefore $D_T \geq O$ (i.e., T is hyponormal).

Solution 7.5. Let T be an arbitrary operator on a Hilbert space $\mathcal{H}$ and consider the following subset of $\mathcal{H}$.

$$\mathcal{M} = \{ x \in \mathcal{H} \colon \|Tx\| = \|T\| \, \|x\| \}.$$

(a) If $T^*Tx = \|T\|^2 x$, then $\|Tx\|^2 = \langle T^*Tx \, ; x \rangle = \|T\|^2 \|x\|^2$. Conversely, if $\|Tx\| = \|T\| \, \|x\|$, then $\langle T^*Tx \, ; \|T\|^2 x \rangle = \|T\|^2 \|Tx\|^2 = \|T\|^4 \|x\|^2$ and hence

$$\begin{aligned}
\|T^*Tx - \|T\|^2 x\|^2 &= \|T^*Tx\|^2 - 2\operatorname{Re}\langle T^*Tx \, ; \|T\|^2 x \rangle + \|T\|^4 \|x\|^2 \\
&= \|T^*Tx\|^2 - \|T\|^4 \|x\|^2 \\
&\leq \left(\|T^*T\|^2 - \|T\|^4 \right) \|x\|^2 = 0.
\end{aligned}$$

Conclusion: $T^*Tx = \|T\|^2 x$ if and only if $\|Tx\| = \|T\| \, \|x\|$.

(b) Thus $\mathcal{M} = \mathcal{N}(\|T\|^2 I - T^*T)$, and so it a closed linear manifold of $\mathcal{H}$.

(c) Suppose T is hyponormal and take an arbitrary $x \in \mathcal{M}$. According to assertion (a) and Problem 7.3,

$$\|T(Tx)\| \leq \|T\| \, \|Tx\| = \|T\|^2 \|x\| = \| \|T\|^2 x\| = \|T^*Tx\| \leq \|T(Tx)\|.$$

Hence $\|T(Tx)\| = \|T\| \, \|Tx\|$ and so $Tx \in \mathcal{M}$. Thus $\mathcal{M}$ is T-invariant.

(d) If T is normal, then both T and T^* are hyponormal so that, according to assertions (b) and (c), $\mathcal{M} = \mathcal{N}(\|T\|^2 I - T^*T) = \mathcal{N}(\|T^*\|^2 I - TT^*)$ is invariant for both T and T^*, and therefore $\mathcal{M}$ reduces T.

Solution 7.6. Take an arbitrary vector u in $\mathcal{M}$, where $\mathcal{M}$ is a subspace of a Hilbert space $\mathcal{H}$, which is invariant for an operator T on $\mathcal{H}$. Problem 4.7 says that $(T|_\mathcal{M})^* = PT^*|_\mathcal{M}$, where P is the orthogonal projection onto $\mathcal{M}$. Therefore, according to Problems 2.7 and 7.3,

$$\|(T|_\mathcal{M})^* u\| = \|PT^*|_\mathcal{M} u\| \leq \|T^*|_\mathcal{M} u\| = \|T^* u\| \leq \|Tu\| = \|T|_\mathcal{M} u\|$$

if T is hyponormal, and hence $T|_\mathcal{M}$ also is hyponormal (Problem 7.3 again). Moreover, if $T|_\mathcal{M}$ is normal, say $T|_\mathcal{M} = N$, then $T = \begin{pmatrix} N & X \\ O & Y \end{pmatrix}$ in $\mathcal{B}[\mathcal{M} \oplus \mathcal{M}^\perp]$ (Problem 4.2). Since N is normal and T is hyponormal, it follows that

$$O \leq D_T = T^*T - TT^* = \begin{pmatrix} -XX^* & N^*X - XY^* \\ X^*N - YX^* & XX^* \end{pmatrix}.$$

With $x = (u, 0)$ in $\mathcal{M} \oplus \mathcal{M}^\perp$ we have $D_T x = (-XX^* u , \; X^* N - YX^* u)$ so that $0 \leq \langle D_T x \, ; x \rangle = \langle -XX^* u \, ; u \rangle = -\|X^* u\|^2$. Then $X^* = O$; that is, $X = O$; and so $T = \left(\begin{smallmatrix} N & O \\ O & Y \end{smallmatrix} \right) = N \oplus Y$. Hence (Problem 4.8) $\mathcal{M}$ reduces T.

Solution 7.7. Let $\mathcal{M}$ be an invariant subspace for an operator T in $\mathcal{B}[\mathcal{H}]$. If T is normal, then it is hyponormal. Therefore, according to Problem 7.6, $\mathcal{M}$ reduces T whenever $T|_\mathcal{M}$ is normal. Conversely, if $\mathcal{M}$ reduces T, then $T = A \oplus D$ on $\mathcal{H} = \mathcal{M} \oplus \mathcal{M}^\perp$, where $A = T|_\mathcal{M}$ and $D = T|_{\mathcal{M}^\perp}$ (Problem 4.8). It is trivially verified that A and D are both normal whenever T is.

Solution 7.8. Take an arbitrary positive integer m. An elementary (and rather intuitive) result of number theory says that every positive integer n can be written as $n = mp_n + q_n$ for some pair (p_n, q_n) of nonnegative integers, where $q_n < m$. Hence

$$\|T^n\| = \|T^{mp_n + q_n}\| = \|T^{mp_n} T^{q_n}\| \leq \|T^{mp_n}\| \|T^{q_n}\| \leq \|T^m\|^{p_n} \|T^{q_n}\|.$$

Put $\mu = \max_{0 \leq k \leq m-1} \|T^k\|$ and recall that $q_n \leq m - 1$. Thus

$$\|T^n\|^{\frac{1}{n}} \leq \|T^m\|^{\frac{p_n}{n}} \mu^{\frac{1}{n}} = \mu^{\frac{1}{n}} \|T^m\|^{\frac{1}{m} - \frac{q_n}{mn}}.$$

Since $\mu^{\frac{1}{n}} \to 1$ and $\|T^m\|^{\frac{1}{m} - \frac{q_n}{mn}} \to \|T^m\|^{\frac{1}{m}}$ as $n \to \infty$, it follows that

$$\limsup_n \|T^n\|^{\frac{1}{n}} \leq \|T^m\|^{\frac{1}{m}}$$

for every positive integer m. Therefore $\limsup_n \|T^n\|^{\frac{1}{n}} \leq \liminf_n \|T^n\|^{\frac{1}{n}}$, which ensures that $\{\|T^n\|^{\frac{1}{n}}\}$ converges in $\mathbb{R}$.

Solution 7.9. Recall that $r(T)^n = r(T^n) \leq \|T^n\| \leq \|T\|^n$ for every $n \geq 0$. Thus $r(T) = \|T\|$ implies $\|T^n\| = \|T\|^n$ for every $n \geq 0$. Conversely, if $\|T^n\| = \|T\|^n$ for every $n \geq 0$, then $r(T) = \lim_n \|T^n\|^{\frac{1}{n}} = \|T\|$.

Solution 7.10. Take $T \in \mathcal{B}[\mathcal{H}]$ and let x be an arbitrary vector in $\mathcal{H}$.

(a) Take an arbitrary integer $n \geq 1$ and note that

$$\|T^n x\|^2 = \langle T^n x \, ; T^n x \rangle = \langle T^* T^n x \, ; T^{n-1} x \rangle \leq \|T^* T^n x\| \|T^{n-1} x\|$$

by the Schwarz inequality. If T is hyponormal, then

$$\|T^* T^n x\| \|T^{n-1} x\| \leq \|T^{n+1} x\| \|T^{n-1} x\| \leq \|T^{n+1}\| \|T^{n-1}\| \|x\|^2$$

by Problem 7.3, and hence

$$\|T^n x\|^2 \leq \|T^{n+1}\| \|T^{n-1}\| \|x\|^2.$$

Thus $\|T^n\|^2 = \sup_{\|x\|=1} \|T^x\|^2 \leq \|T^{n+1}\| \|T^{n-1}\|$, which proves (a).

(b) $\|T^n\| = \|T\|^n$ holds trivially if $T = O$ (for all $n \geq 0$) and if $n = 0, 1$ (for all T in $\mathcal{B}[\mathcal{H}]$). Let T be a nonzero operator and suppose $\|T^n\| = \|T\|^n$ for some integer $n \geq 1$. If $\|T^n\|^2 \leq \|T^{n+1}\|\|T^{n-1}\|$, then

$$\|T\|^{2n} = (\|T\|^n)^2 = \|T^n\|^2 \leq \|T^{n+1}\|\|T^{n-1}\| \leq \|T^{n+1}\|\|T\|^{n-1}$$

since $\|T^m\| \leq \|T\|^m$ for every $m \geq 0$, and therefore (recall: $T \neq O$),

$$\|T\|^{n+1} = \|T\|^{2n}(\|T\|^{n-1})^{-1} \leq \|T^{n+1}\| \leq \|T\|^{n+1}.$$

Hence $\|T^{n+1}\| = \|T\|^{n+1}$, concluding the proof of (b) by induction.

(c) Assertions (a), (b) and Problem 7.9 ensure (c).

Solution 7.11. Let $\mathcal{H}$ and $\mathcal{K}$ be unitarily equivalent Hilbert spaces. Take any operator T in $\mathcal{B}[\mathcal{H}]$ and any unitary transformation U in $\mathcal{B}[\mathcal{K}, \mathcal{H}]$. Set $S = U^*TU$ in $\mathcal{B}[\mathcal{K}]$ so that $S^* = U^*T^*U$. The self-commutators are related by $D_S = U^*D_TU$, and therefore S is normal if and only if T is. Now let $\mathcal{Z}$ be any Hilbert space, take $N_1 : \mathcal{H} \oplus \mathcal{Z} \to \mathcal{H} \oplus \mathcal{Z}$ and $N_2 : \mathcal{K} \oplus \mathcal{Z} \to \mathcal{K} \oplus \mathcal{Z}$ such that

$$N_1 = \begin{pmatrix} T & X \\ O & Y \end{pmatrix} \quad \text{and} \quad N_2 = \begin{pmatrix} S & U^*X \\ O & Y \end{pmatrix}, \quad \text{and set} \quad W = \begin{pmatrix} U & O \\ O & I \end{pmatrix};$$

with $X : \mathcal{Z} \to \mathcal{H}$, $Y : \mathcal{Z} \to \mathcal{Z}$ and $I : \mathcal{Z} \to \mathcal{Z}$, the identity on $\mathcal{Z}$. Since $N_2 = W^*N_1W$ and since W is unitary, it follows that if one of N_1 or N_2 is normal, then so is the other. Thus S subnormal if and only if T is subnormal.

Solution 7.12. Take a unilateral weighted shift with nonnegative weights $T_+ = \text{shift}(\{\alpha_k\}_{k=0}^\infty)$. Put $\beta_n = \prod_{k=0}^{n-1} \alpha_k$ so that $\alpha_n\beta_n\beta_{n+2} = \beta_{n+1}^2\alpha_{n+1}$, for every $n \geq 1$. If $\alpha_m = \alpha_{m+1}$ for some $m \geq 1$, then $\beta_m\beta_{m+2} = \beta_{m+1}^2$. If T_+ is injective, subnormal and $\|T_+\| = 1$, then there exists a probability measure ν on $[0, 1]$ containing 1 in its support such that $\beta_n^2 = \int x^{2n} d\nu(x)$ for each $n \geq 1$. Let $L^2(\nu)$ be the Hilbert space of all square-summable functions on $[0, 1]$ with respect to ν. Take p_n in $L^2(\nu)$ such that $p_n(x) = x^n$ (equality in the sense of $L^2(\nu)$; that is, $p_n(x) = x^n$ almost everywhere in $[0, 1]$ with respect to ν) so that $\|p_n\|^2 = \int x^{2n} d\nu(x) = \beta_n^2$ for each $n \geq 1$. Thus

$$\langle p_m ; p_{m+2} \rangle = \int x^m x^{m+2} d\nu(x) = \int x^{2(m+1)} d\nu(x) = \beta_{m+1}^2 = \beta_m\beta_{m+2}$$

$$= \left(\int x^{2m} d\nu(x) \right)^{\frac{1}{2}} \left(\int x^{2(m+2)} d\nu(x) \right)^{\frac{1}{2}} = \|p_m\|\|p_{m+2}\|$$

and the Schwartz inequality becomes an identity. This means that p_m and p_{m+2} are collinear; that is (since they are not null), $p_m = \gamma p_{m+2}$ for some nonzero constant γ so that $x^m - \gamma x^{m+2} = 0$ almost everywhere on $[0, 1]$ with respect to ν. Since $x \mapsto x^m - \gamma x^{m+2}$ is a continuous function on $[0, 1]$, this implies that $x^m - \gamma x^{m+2} = 0$ everywhere on support ν and so it holds at $x = 1$ because 1 lies in support ν. Therefore $\gamma = 1$ so that $x^m = x^{m+2}$ everywhere on support ν, and hence support ν is included in $\{0,1\}$ (since $x^m \neq x^{m+2}$ for every $x \in (0,1)$); that is, $1 \in [\nu] \subseteq \{0,1\}$, where $[\nu]$ stands for support ν. Then $\beta_n^2 = \int x^{2n} d\nu(x) = \beta_1^2 = \alpha_0^2 \neq 0$ for all $n \geq 1$. But we know that $\beta_{n+1} = \beta_n \alpha_n$, and so $\alpha_n = 1$ for all $n \geq 1$. That is,

$$T_+ = \text{shift}(\alpha_0, 1, 1, 1, \ldots).$$

Conversely, if $\alpha = 1$, then $T_+ = \text{shift}(\alpha, 1, 1, 1, \ldots) = S_+$ is quasinormal (S_+ is an isometry), and hence subnormal. This is the trivial case. Now take any α in $(0,1)$. Note that $T_+ = \text{shift}(\alpha, 1, 1, 1, \ldots)$ is injective and $\|T_+\| = 1$. Let δ_0 and δ_1 be unit point measures concentrated at 0 and 1, respectively (i.e., Dirac measures). Set $\nu = (1 - \alpha^2)\delta_0 + \alpha^2 \delta_1$, a probability measure on $[0,1]$ containing 1 in its support (in fact, $[\nu] = \{0,1\}$ for $0 < \alpha < 1$ and $\int d\nu(x) = 1$). Then $\prod_{k=0}^{n-1} \alpha_k^2 = \int x^{2n} d\nu(x) = \alpha^2$ for all $n \geq 1$, and therefore T_+ is subnormal.

Solution 7.13. If $0 < \alpha_k \leq 1 = \sup_{k \geq 0} |\alpha_k|$ for every $k \geq 0$, then T_+ is injective (with positive weights) and $\|T_+\| = 1$. If T_+ is subnormal, then there exists a probability measure ν on $[0,1]$ containing 1 in its support such that $\prod_{k=0}^{n-1} \alpha_k^2 = \int x^{2n} d\nu(x)$ for every $n \geq 1$. Suppose $\alpha_0 = \alpha_1$. In this case we get $\int x^4 d\nu(x) = (\int x^2 d\nu(x))^2 = \alpha_0^4$ so that

$$\int (x^2 - \alpha_0^2)^2 d\nu(x) = \int x^4 d\nu(x) - \alpha_0^4 = 0$$

(i.e., variance of x^2 is null). Thus $x^2 = \alpha_0^2$ almost everywhere on $[0,1]$ with respect to ν. Since $x \mapsto x^2 - \alpha_0^2$ is a continuous function on $[0,1]$, it follows that $x^2 = \alpha_0^2$ everywhere on support ν. This implies that ν is a unit point measure concentrated at α_0^2. But a unit point measure containing 1 in its support is concentrated at 1. Hence $\alpha_0^2 = 1$. Outcome: if $\alpha_0^2 \neq 1$ (that is, if $\alpha_0 \neq 1$), then T_+ is not subnormal.

Solution 7.14. If $\alpha_k = \left(\frac{k+1}{k+2}\right)^{\frac{1}{2}}$ so that $\alpha_k^2 = 1 - \frac{1}{k+2}$, for every $k \geq 0$, then $T_+ = \text{shift}(\{\alpha_k\}_{k=0}^\infty)$ is injective and $\|T_+\| = 1$ (the positive weight sequence $\{\alpha_k\}_{k=0}^\infty$ is increasing and $\lim_k \alpha_k = 1$). Consider the measure ν on $[0,1]$ given by $d\nu(x) = 2x \, d\mu(x)$, where μ is the Lebesgue measure on $[0,1]$. Put dx for $d\mu(x)$, as usual. It is clear that ν is a probability measure on $[0,1]$ containing 1 in its support (indeed, $[\nu] = [0,1]$ and $\int d\nu(x) = 2\int x \, dx = 1$). It is readily verified by induction that $\prod_{k=0}^{n-1} \alpha_k^2 = \frac{1}{n+1}$ for each $n \geq 1$. Thus

$$\int x^{2n} d\nu(x) = \int x^{2n} 2x \, dx = \frac{1}{n+1} = \prod_{k=0}^{n-1} \alpha_k^2$$

for every $n \geq 1$, and hence T_+ is subnormal.

8
Spectral Properties

One of the fundamental theorems of (linear) functional analysis, namely, the Open Mapping Theorem, says that *a continuous linear transformation of a Banach space onto a Banach space is an open mapping* (a mapping is *open* if it maps open sets onto open sets). A crucial corollary is the Inverse Mapping Theorem (see e.g., [32, p. 228]).

Inverse Mapping Theorem. *If $\mathcal{X}$ and $\mathcal{Y}$ are Banach spaces and if T in $\mathcal{B}[\mathcal{X}, \mathcal{Y}]$ is injective and surjective, then T^{-1} lies in $\mathcal{B}[\mathcal{Y}, \mathcal{X}]$.*

In other words, every injective and surjective bounded linear transformation between Banach spaces has a *bounded* (linear) inverse. Let $\mathcal{G}[\mathcal{X}, \mathcal{Y}]$ denote the collection of all invertible elements from $\mathcal{B}[\mathcal{X}, \mathcal{Y}]$. The above theorem says that, if $\mathcal{X}$ and $\mathcal{Y}$ are Banach spaces, then $T^{-1} \in \mathcal{G}[\mathcal{Y}, \mathcal{X}]$ whenever $T \in \mathcal{G}[\mathcal{X}, \mathcal{Y}]$. Moreover, the inverse of the product of invertible bounded linear transformations is again a bounded linear transformation. Precisely, if $T \in \mathcal{G}[\mathcal{X}, \mathcal{Y}]$ and $S \in \mathcal{G}[\mathcal{Y}, \mathcal{Z}]$, where $\mathcal{X}$, $\mathcal{Y}$ and $\mathcal{Z}$ are Banach spaces, then $ST \in \mathcal{G}[\mathcal{X}, \mathcal{Z}]$ (for $\|ST\| \leq \|S\| \|T\|$ and $(ST)^{-1} = T^{-1}S^{-1}$ so that $\|(ST)^{-1}\| \leq \|T^{-1}\| \|S^{-1}\|$). Set $\mathcal{G}[\mathcal{X}] = \mathcal{G}[\mathcal{X}, \mathcal{X}]$. Thus $\mathcal{G}[\mathcal{X}]$ forms a group under multiplication whenever $\mathcal{X}$ is a Banach space, namely, the group of all invertible operators from $\mathcal{B}[\mathcal{X}]$ (every operator in $\mathcal{G}[\mathcal{X}]$ has an inverse in $\mathcal{G}[\mathcal{X}]$). Here is a useful corollary of the Inverse Mapping Theorem, which we shall baptize Bounded Inverse Theorem.

Bounded Inverse Theorem. *Let $\mathcal{X}$ and $\mathcal{Y}$ be Banach spaces and take any T in $\mathcal{B}[\mathcal{X}, \mathcal{Y}]$. The following assertions are pairwise equivalent.*

(a) *There exists $T^{-1} \in B[\mathcal{R}(T), \mathcal{X}]$*   (i.e., *T has a bounded inverse on its range*).

(b) *T is bounded below*   (i.e., *there exists a constant $\alpha > 0$ such that $\alpha\|x\| \le \|Tx\|$ for every $x \in \mathcal{X}$*).

(c) $\mathcal{N}(T) = \{0\}$ *and* $\mathcal{R}(T)^- = \mathcal{R}(T)$ (i.e., *T is injective and has a closed range*).

Remarks: Recall that a linear transformation is injective if and only if it has a null kernel. The equivalence between assertions (a) and (b) still hold if $\mathcal{X}$ and $\mathcal{Y}$ are just normed spaces. If $\mathcal{X}$ is a Banach space, then each of them implies (c). A trivial (but important) consequence is: *every isometry on a Banach space has a closed range*. The role played by the Inverse Mapping Theorem in the proof of the above theorem is restricted to show that (c) implies (a), the rest of the proof of the Bounded Inverse Theorem does not need the Inverse Mapping Theorem (see e.g., [32, pp. 223,228]).

Let $\mathcal{X}$ be a *complex* Banach space and consider the Banach algebra $B[\mathcal{X}]$ of all operators on $\mathcal{X}$. Take any T in $B[\mathcal{X}]$. The set $\rho(T)$ of all complex numbers λ for which $(\lambda I - T)$ is invertible is the *resolvent set* of T,

$$\rho(T) = \{\lambda \in \mathbb{C}\colon (\lambda I - T) \in \mathcal{G}[\mathcal{X}]\}$$
$$= \{\lambda \in \mathbb{C}\colon \mathcal{N}(\lambda I - T) = \{0\} \text{ and } \mathcal{R}(\lambda I - T) = \mathcal{X}\}.$$

The complement of $\rho(T)$, denoted by $\sigma(T)$, is the *spectrum* of T. Thus

$$\sigma(T) = \mathbb{C}\backslash\rho(T) = \{\lambda \in \mathbb{C}\colon \mathcal{N}(\lambda I - T) \neq \{0\} \text{ or } \mathcal{R}(\lambda I - T) \neq \mathcal{X}\}.$$

A basic property of the spectrum of an operator on a complex Banach space is that it is a *nonempty compact* subset of $\mathbb{C}$ (i.e., $\sigma(T) \neq \varnothing$ is closed and bounded in $\mathbb{C}$). This happens because $\mathcal{X}$ is complex and T is bounded (see e.g., [32, pp. 449,450]). As the complement of $\rho(T)$, the spectrum $\sigma(T)$ is precisely the set of all λ in $\mathbb{C}$ such that the operator $(\lambda I - T)$ fails to be invertible (i.e., fails to have a bounded inverse defined on the whole space $\mathcal{X}$). According to the origin of such a failure the spectrum can be split into many disjoint parts. A classical partition comprises three parts. The set of those λ such that $(\lambda I - T)$ has no inverse is the *point spectrum* of T:

$$\sigma_P(T) = \{\lambda \in \mathbb{C}\colon \mathcal{N}(\lambda I - T) \neq \{0\}\},$$

which is exactly the set of all eigenvalues of T. Recall that a scalar $\lambda \in \mathbb{C}$ is an *eigenvalue* of an operator T if there exists a nonzero vector x in $\mathcal{X}$ such that $Tx = \lambda x$. Equivalently, if $\mathcal{N}(\lambda I - T) \neq \{0\}$. If λ is an eigenvalue of T, then the nonzero vectors in $\mathcal{N}(\lambda I - T)$ are the *eigenvectors* of T associated with λ, and the subspace $\mathcal{N}(\lambda I - T)$ of $\mathcal{X}$ is the *eigenspace* of T associated with λ. The *multiplicity* of an eigenvalue is the dimension of the respective eigenspace (i.e., $\dim \mathcal{N}(\lambda I - T)$). The set of those λ for which $(\lambda I - T)$ has a densely defined but unbounded inverse on its range is the *continuous spectrum* of T:

$$\sigma_C(T) = \{\lambda \in \mathbb{C}: \mathcal{N}(\lambda I - T) = \{0\}, \mathcal{R}(\lambda I - T)^- = \mathcal{X} \text{ and } \mathcal{R}(\lambda I - T) \neq \mathcal{X}\}.$$

If $(\lambda I - T)$ has an inverse on its range that is not densely defined, then λ belongs to the *residual spectrum* T:

$$\sigma_R(T) = \{\lambda \in \mathbb{C}: \mathcal{N}(\lambda I - T) = \{0\} \text{ and } \mathcal{R}(\lambda I - T)^- \neq \mathcal{X}\}.$$

Note that, if $(\lambda I - T)$ is bounded below, then the Bounded Inverse Theorem ensures that $\lambda \in \rho(T) \cup \sigma_R(T)$. The collection $\{\sigma_P(T), \sigma_C(T), \sigma_R(T)\}$ forms a partition of $\sigma(T)$; that is, they are pairwise disjoint and

$$\sigma(T) = \sigma_P(T) \cup \sigma_C(T) \cup \sigma_R(T).$$

In a Hilbert space setting we get a slight simplification that allows us to characterize all the above parts of the spectrum once we know the spectrum $\sigma(T)$ itself and the point spectra $\sigma_P(T)$ and $\sigma_P(T^*)$, as we shall see next. But first we need the following notation. If Λ is any subset of $\mathbb{C}$, then let $\Lambda^* = \{\bar{\lambda} \in \mathbb{C}: \lambda \in \Lambda\}$ be the set of the complex conjugates of the elements of Λ. That is, Λ^* is the mirror image of Λ with respect to the real axis. It is clear that $\Lambda^{**} = \Lambda$ and $(\mathbb{C}\backslash\Lambda)^* = \mathbb{C}\backslash\Lambda^*$.

Problem 8.1. If $T \in \mathcal{B}[\mathcal{H}]$, where $\mathcal{H}$ is a complex Hilbert space, then

$$\rho(T) = \rho(T^*)^*, \quad \sigma(T) = \sigma(T^*)^*, \quad \sigma_C(T) = \sigma_C(T^*)^*.$$

Also show that the residual spectrum is given by the formula

$$\sigma_R(T) = \sigma_P(T^*)^* \backslash \sigma_P(T).$$

Therefore,

$$\sigma_C(T) = \sigma(T)\backslash\big(\sigma_P(T) \cup \sigma_R(T)\big) = \sigma(T)\backslash\big(\sigma_P(T) \cup \sigma_P(T^*)^*\big).$$

Hint: Take any $S \in \mathcal{B}[\mathcal{H}]$. $\mathcal{R}(S)^- = \mathcal{R}(S)$ if and only if $\mathcal{R}(S^*)^- = \mathcal{R}(S^*)$, and $S \in \mathcal{G}[\mathcal{H}]$ if and only if $S^* \in \mathcal{G}[\mathcal{H}]$ (see e.g., [32, pp. 389,394]).

Problem 8.2. Let T be an operator acting on a complex normed space $\mathcal{X}$. Prove the following assertions.

(a) For each $\lambda \in \mathbb{C}$ the subspace $\mathcal{N}(\lambda I - T)$ is hyperinvariant for T.

(b) Every eigenspace of a nonscalar operator is a *nontrivial* hyperinvariant subspace for it.

(c) If T has no nontrivial invariant subspace and $\mathcal{X}$ has dimension greater than 1, then $\sigma_P(T) = \varnothing$.

Now suppose T acts on a complex Hilbert space $\mathcal{H}$.

(d) If T is nonscalar and $\sigma_P(T) \cup \sigma_R(T) \neq \varnothing$ (i.e., $\sigma_P(T) \cup \sigma_P(T^*) \neq \varnothing$), then T has a nontrivial hyperinvariant subspace.

(e) If T has no nontrivial invariant subspace and $\mathcal{H}$ has dimension greater than 1, then $\sigma(T) = \sigma_C(T)$.

Problem 8.3. Let $\mathcal{H} \neq \{0\}$ be a complex Hilbert space and let Γ be the unit circle about the origin of the complex plane. Show that

(a) if $H \in \mathcal{B}[\mathcal{H}]$ is hyponormal, then $\sigma_P(H)^* \subseteq \sigma_P(H^*)$ and $\sigma_R(H^*) = \varnothing$;

(b) if $N \in \mathcal{B}[\mathcal{H}]$ is normal, then $\sigma_P(N^*) = \sigma_P(N)^*$ and $\sigma_R(N) = \varnothing$;

(c) if $U \in \mathcal{B}[\mathcal{H}]$ is unitary, then $\sigma(U) \subseteq \Gamma$;

(d) if $A \in \mathcal{B}[\mathcal{H}]$ is self-adjoint, then $\sigma(A) \subset \mathbb{R}$;

(e) if $Q \in \mathcal{B}[\mathcal{H}]$ is nonnegative, then $\sigma(Q) \subset [0, \infty)$;

(f) if $R \in \mathcal{B}[\mathcal{H}]$ is strictly positive, then $\sigma(R) \subset [\alpha, \infty)$ for some $\alpha > 0$;

(g) if $P \in \mathcal{B}[\mathcal{H}]$ is a nontrivial projection, then $\sigma(P) = \sigma_P(P) = \{0, 1\}$.

Let $T \in \mathcal{B}[\mathcal{X}]$ be an operator on a complex Banach space $\mathcal{X}$. The *spectral radius* of T is the number

$$r_\sigma(T) \;=\; \sup_{\lambda \in \sigma(T)} |\lambda| \;=\; \max_{\lambda \in \sigma(T)} |\lambda|.$$

The first identity defines the spectral radius; the second one follows from the Weierstrass Theorem: *A continuous real-valued function on a metric space assumes both a maximum and minimum value on each compact set.* It is easy to verify (see e.g., [32, p. 459]) that

$$r_\sigma(T^n) = r_\sigma(T)^n \leq \|T^n\| \leq \|T\|^n \quad \text{for every} \quad n \geq 0.$$

An operator T is *quasinilpotent* if $r_\sigma(T) = 0$ (i.e., $\sigma(T) = \{0\}$). The above expression shows at once that *every nilpotent is quasinilpotent*, and also that *if T is power bounded, then $r_\sigma(T) \leq 1$*. Moreover, asymptotically, the first inequality leads to the Gelfand–Beurling formula (see e.g., [32, p. 460]).

Gelfand–Beurling Formula. $r_\sigma(T) = \lim_n \|T^n\|^{\frac{1}{n}}$.

What the Gelfand–Beurling formula says is that $r_\sigma(T) = r(T)$, where $r(T)$ is the limit of the sequence $\{\|T^n\|^{\frac{1}{n}}\}$ of Problem 7.8. We adopt the simplest notation and, from now on, write $r(T)$ for the spectral radius:

$$r(T) \;=\; \sup_{\lambda \in \sigma(T)} |\lambda| \;=\; \max_{\lambda \in \sigma(T)} |\lambda| \;=\; \lim_n \|T^n\|^{\frac{1}{n}}.$$

Therefore, *a normaloid operator acting on a complex Banach space is precisely an operator whose norm coincides with the spectral radius.* Note that $r(\alpha T) = |\alpha| r(T)$ for every scalar α. In a complex Hilbert space $\mathcal{H}$ every normal operator is normaloid, and so is every nonnegative operator. Since T^*T is always nonnegative, it follows that, for every T in $\mathcal{B}[\mathcal{H}]$,

$$r(T^*T) = r(TT^*) = \|T^*T\| = \|TT^*\| = \|T\|^2 = \|T^*\|^2.$$

Recall from Chapter 7 that $r(T^*) = r(T)$; indeed, $\sigma(T^*) = \sigma(T)^*$ (Problem 8.1). Note that T *is normaloid if and only if there exists* λ *in* $\sigma(T)$ *such that* $|\lambda| = \|T\|$. However, such a λ can never be in the residual spectrum.

Problem 8.4. Show that $\sigma_R(T) \subseteq \{\lambda \in \mathbb{C} : |\lambda| < \|T\|\}$ for every $T \in \mathcal{B}[\mathcal{H}]$.

Use the Gelfand–Beurling formula to compute the spectral radius of a unilateral weighted shift $T_+ = \text{shift}(\{\alpha_k\}_{k=0}^\infty)$ on ℓ_+^2.

Problem 8.5. Show that $r(T_+) = \lim_n \sup_k \left(\prod_{j=k}^{n+k-1} |\alpha_j| \right)^{\frac{1}{n}}$.

A significant application of the Gelfand–Beurling formula ensures that *an operator T is uniformly stable if and only if $r(T) < 1$.*

Problem 8.6. Let $T \in \mathcal{B}[\mathcal{X}]$ be an operator on a complex Banach space $\mathcal{X}$. Show that the following assertions are pairwise equivalent.

(a) $T^n \xrightarrow{u} O$.

(b) $r(T) < 1$.

(c) $\|T^n\| \leq \beta \alpha^n$ for every $n \geq 0$, for some $\beta \geq 1$ and some $\alpha \in (0, 1)$.

Recall that a set (in a topological space; in particular, in a metric space) is *compact* if every covering of it made up of open sets has a finite subcovering. In a metric space a set is compact if and only if it is *sequentially compact:* every sequence whose entries lie in the set has a subsequence that converges in it. A linear transformation between normed spaces is *compact* if the closure of the image of each bounded set is a compact set, which means that it maps bounded sequences into sequences that have a convergent subsequence. In a Hilbert space setting this is equivalent to saying that it maps weakly convergent sequences into convergent sequences. Compact linear transformations are bounded (i.e., continuous), and the adjoint of a compact linear transformation between Hilbert spaces is again a compact linear transformation. Let $\mathcal{B}_\infty[\mathcal{X}, \mathcal{Y}]$ denote the collection of all compact linear transformations of a normed space $\mathcal{X}$ into a normed space $\mathcal{Y}$. Thus $\mathcal{B}_\infty[\mathcal{X}, \mathcal{Y}] \subseteq \mathcal{B}[\mathcal{X}, \mathcal{Y}]$ (if $\mathcal{X}$ is finite-dimensional, then $\mathcal{B}_\infty[\mathcal{X}, \mathcal{Y}] = \mathcal{B}[\mathcal{X}, \mathcal{Y}]$). In fact, $\mathcal{B}_\infty[\mathcal{X}, \mathcal{Y}]$ is a linear manifold of $\mathcal{B}[\mathcal{X}, \mathcal{Y}]$ and, if $\mathcal{Y}$ is a Banach space, then $\mathcal{B}_\infty[\mathcal{X}, \mathcal{Y}]$ is a subspace (i.e., a closed linear manifold) of $\mathcal{B}[\mathcal{X}, \mathcal{Y}]$. Put

$\mathcal{B}_\infty[\mathcal{X}] = \mathcal{B}_\infty[\mathcal{X}, \mathcal{X}]$ so that $\mathcal{B}_\infty[\mathcal{X}] \subseteq \mathcal{B}[\mathcal{X}]$. Actually, $\mathcal{B}_\infty[\mathcal{X}]$ is a *two-sided ideal* of the algebra $\mathcal{B}[\mathcal{X}]$, which means that $\mathcal{B}_\infty[\mathcal{X}]$ is a subalgebra of $\mathcal{B}[\mathcal{X}]$ such that the product (both left and right products) of every element of $\mathcal{B}_\infty[\mathcal{X}]$ with any element of $\mathcal{B}[\mathcal{X}]$ is again an element of $\mathcal{B}_\infty[\mathcal{X}]$. We shall summarize below the spectral properties of compact operators that will be required in the sequel (see e.g., [8, §VII.7] or [32, §6.6]).

Fredholm Alternative. *If T is a compact operator on a complex Banach space $\mathcal{X}$, then*

$$\sigma(T) \backslash \{0\} = \sigma_P(T) \backslash \{0\}.$$

Moreover, $\sigma(T)$ is countable and 0 is the only possible accumulation point of it. If $\mathcal{X}$ is a Hilbert space, then $\dim \mathcal{N}(\lambda I - T) = \dim \mathcal{N}(\bar{\lambda} I - T^*) < \infty$.

The scalar 0 may be anywhere. However, if T is a compact operator on a complex Banach space $\mathcal{X}$ and $0 \in \rho(T)$, then $\mathcal{X}$ must be finite-dimensional. Indeed, if $0 \in \rho(T)$, then $T^{-1} \in \mathcal{B}[\mathcal{X}]$ so that $I = T^{-1}T$ is compact, which implies that $\mathcal{X}$ is finite-dimensional (reason: the identity I is not compact on an infinite-dimensional space). The Fredholm alternative will be applied often from now on. Here is a first application. Let $\mathcal{B}_0[\mathcal{X}]$ denote the class of all finite-rank operators on $\mathcal{X}$ (operators with a finite-dimensional range). It is easy to verify that $\mathcal{B}_0[\mathcal{X}] \subseteq \mathcal{B}_\infty[\mathcal{X}]$ (finite-rank operators are compact).

Problem 8.7. If $T \in \mathcal{B}_0[\mathcal{X}]$, then the spectrum of T is finite and coincides with the point spectrum. That is, if T is a finite-rank operator on a complex Banach space $\mathcal{X}$ (in particular, if $\mathcal{X}$ is finite-dimensional), then show that

$$\sigma(T) = \sigma_P(T), \quad \text{which is finite.}$$

Hint: Recall that $\dim \mathcal{X} = \dim \mathcal{R}(L) + \dim \mathcal{N}(L)$ for every linear transformation $L: \mathcal{X} \to \mathcal{Y}$ between linear spaces $\mathcal{X}$ and $\mathcal{Y}$. Also recall that eigenvectors associated with distinct eigenvalues are linearly independent.

Problem 8.8. Let T be an operator on a complex Hilbert space and let Δ be the open unit disc about the origin of the complex plane. Show that

(a) $T^n \xrightarrow{w} O$ implies $\sigma_P(T) \subseteq \Delta$.

If T is a compact operator, then use the Fredholm alternative and the Gelfand–Beurling formula to show that

(b) $\sigma_P(T) \subseteq \Delta$ implies $T^n \xrightarrow{u} O$.

(c) Thus conclude: The concepts of weak, strong, and uniform stabilities coincide for a compact operator on a complex Hilbert space.

The next problem extends the previous one. *If the continuous spectrum does not intersect the unit circle, then the concepts of weak, strong, and uniform stabilities coincide.* (The continuous spectrum of a compact operator T is a subset of $\{0\}$ — Fredholm alternative; and empty if T is finite-rank).

Problem 8.9. Take an operator T on a complex Hilbert space and let Γ be the unit circle (about the origin of the complex plane). Show that

$$T^n \xrightarrow{u} O \quad \text{if and only if} \quad T^n \xrightarrow{w} O \text{ and } \sigma_C(T) \cap \Gamma = \varnothing.$$

Consider the result of Problem 6.6. It is clear that $A = A_*$ for every normal contraction. (Indeed, if T is normal, then $T^{*n}T^n = T^nT^{*n}$ for every $n \geq 1$.) Now use Problem 8.8 to show that this is also the case for every compact contraction on a complex Hilbert space.

Problem 8.10. If T is a compact contraction, then $A = A_*$.

The Spectral Theorem is a milestone in the theory of Hilbert space operators, providing a full statement about the nature and structure of normal operators. It is usual to say that it can be applied to answer essentially all questions on normal operators, which indeed is the case as far as "essentially all" means "almost all" or "all the principal"; there exist open questions on normal operators. For compact normal operators the Spectral Theorem does not require any knowledge of measure theory, yielding the concept of diagonalization (see e.g., [32, §6.7]). However, in its full version, it requires the notion of spectral measure. Let Ω be a set in the complex plane $\mathbb{C}$ and let Σ_Ω be the σ-algebra of Borel sets in Ω. A (complex) *spectral measure* in a (complex) Hilbert space $\mathcal{H}$ is a mapping $P \colon \Sigma_\Omega \to \mathcal{B}[\mathcal{H}]$ such that

(a) $P(\Lambda)$ is an orthogonal projection for every $\Lambda \in \Sigma_\Omega$,

(b) $P(\varnothing) = O$ and $P(\Omega) = I$,

(c) $P(\Lambda_1 \cap \Lambda_2) = P(\Lambda_1)P(\Lambda_2)$ for every $\Lambda_1, \Lambda_2 \in \Sigma_\Omega$,

(d) if $\{\Lambda_k\}$ is a countable collection of pairwise disjoint sets in Σ_Ω, then

$$P\left(\bigcup_k \Lambda_k\right) = \sum_k P(\Lambda_k),$$

where the series converges strongly if the collection $\{\Lambda_k\}$ is infinite.

Property (c) ensures that the orthogonal projections $P(\Lambda_1)$ and $P(\Lambda_2)$ are orthogonal to each other whenever Λ_1 and Λ_2 are disjoint sets in Σ_Ω, while property (d) says that $\sum_{k=1}^n P(\Lambda_k) \xrightarrow{s} P\left(\bigcup_{k=1}^\infty \Lambda_k\right)$ when $\{\Lambda_k\}_{k \in \mathbb{N}}$ is a sequence of pairwise disjoint sets in Σ_Ω. If $\{\Lambda_k\}_{k \in \mathbb{N}}$ is a partition of Ω, then $\sum_{k=1}^n P(\Lambda_k) \xrightarrow{s} P(\Omega) = I$. Now take a pair of vectors $x, y \in \mathcal{H}$ and consider the mapping $p_{x,y} \colon \Sigma_\Omega \to \mathbb{C}$ defined by

$$p_{x,y}(\Lambda) = \langle P(\Lambda)x \, ; y \rangle \quad \text{for every} \quad \Lambda \in \Sigma_\Omega.$$

This is an ordinary complex-valued countably additive measure on Σ_Ω for each pair $x, y \in \mathcal{H}$. Let $\varphi \colon \Omega \to \mathbb{C}$ be an arbitrary bounded Σ_Ω-measurable function. The integral of φ with respect to $p_{x,y}$, namely, $\int \varphi(\lambda) \, dp_{x,y}(\lambda)$, will be denoted by $\int \varphi(\lambda) \, d\langle P_\lambda x \, ; y \rangle$. It can be shown that, for each function φ as above, there exists a unique operator $F \in \mathcal{B}[\mathcal{H}]$ such that

$$\langle Fx\,;y\rangle = \int \varphi(\lambda)\,d\langle P_\lambda x\,;y\rangle$$

for every $x, y \in \mathcal{H}$ (see e.g., [32, p. 491]). This is usually written as

$$F = \int \varphi(\lambda)\,dP_\lambda,$$

which is just short for $\langle Fx\,;y\rangle = \int \varphi(\lambda)\,d\langle P_\lambda x\,;y\rangle$ for every $x, y \in \mathcal{H}$. If $\psi\colon \Omega \to \mathbb{C}$ is a bounded Σ_Ω-measurable function and $G = \int \psi(\lambda)\,dP_\lambda$, then it can be shown that $FG = \int \varphi(\lambda)\psi(\lambda)\,dP_\lambda$ (see e.g., [47, p. 17]). Take any set Λ in Σ_Ω and let $\chi_\Lambda\colon \Omega \to \mathbb{C}$ be the characteristic function of Λ (which is bounded and Σ_Ω-measurable). Put $\int_\Lambda \varphi(\lambda)\,dP_\lambda = \int \chi_\Lambda(\lambda)\varphi(\lambda)\,dP_\lambda$. Since

$$P(\Lambda) = \int \chi_\Lambda(\lambda)\,dP_\lambda = \int_\Lambda dP_\lambda$$

(reason: $\int \chi_\Lambda(\lambda)\,d\langle P_\lambda x\,;y\rangle = \int_\Lambda d\langle P_\lambda x\,;y\rangle = \langle P(\Lambda)x\,;y\rangle$), we get

$$F|_{\mathcal{R}(P(\Lambda))} = FP(\Lambda) = P(\Lambda)F = \int \chi_\Lambda(\lambda)\varphi(\lambda)\,dP_\lambda = \int_\Lambda \varphi(\lambda)\,dP_\lambda$$

so that each $\mathcal{R}(P(\Lambda))$ is a reducing subspace for F (Problem 4.6). Moreover, since $\bar{p}_{y,x}(\Lambda) = p_{x,y}(\Lambda)$ for every $\Lambda \in \Sigma_\Omega$ and every $x, y \in \mathcal{H}$, it is readily verified that $\langle F^*x\,;y\rangle = \int \bar{\varphi}(\lambda)\,d\langle P_\lambda x\,;y\rangle$ for all $x, y \in \mathcal{H}$. Hence

$$F^* = \int \bar{\varphi}(\lambda)\,dP_\lambda.$$

Thus $F^*F = \int \bar{\varphi}(\lambda)\varphi(\lambda)\,dP_\lambda$ and $FF^* = \int \varphi(\lambda)\bar{\varphi}(\lambda)\,dP_\lambda$ so that

$$F^*F = \int |\varphi(\lambda)|^2 dP_\lambda = FF^*;$$

that is, F is normal. The Spectral Theorem states the converse.

Spectral Theorem. *If $N \in \mathcal{B}[\mathcal{H}]$ is normal, then there exists a unique spectral measure P on $\Sigma_{\sigma(N)}$ such that*

$$N = \int \lambda\,dP_\lambda.$$

If Λ is a nonempty relatively open subset of $\sigma(N)$, then $P(\Lambda) \neq O$.

For a proof of the Spectral Theorem the reader is referred to [8, p. 263] or [10, p. 47] (see also [19, §44]). The representation $N = \int \lambda\,dP_\lambda$ is called the *spectral decomposition* of N. Observe that $N^*N = \int |\lambda|^2\,dP_\lambda = NN^*$.

Problem 8.11. If T is a normal operator, then show that

(a) T is unitary if and only if $\sigma(T) \subseteq \Gamma$,

(b) T is self-adjoint if and only if $\sigma(T) \subset \mathbb{R}$,

(c) T is nonnegative if and only if $\sigma(T) \subset [0, \infty)$,

(d) T is strictly positive if and only if $\sigma(T) \subset [\alpha, \infty)$ for some $\alpha > 0$,

(e) T is an orthogonal projection if and only if $\sigma(T) \subseteq \{0, 1\}$.

Let Q and R be nonnegative operators on a Hilbert space $\mathcal{H}$ (i.e., take Q and R in $\mathcal{B}^+[\mathcal{H}]$). It is easy to prove, by the Square Root Theorem, that

$$\text{if} \quad Q \leq R \quad \text{and} \quad QR = RQ, \quad \text{then} \quad Q^2 \leq R^2.$$

Indeed, if Q and R commute, then $R^2 - Q^2 = (R + Q)(R - Q)$ and $R - Q$ commutes with $R + Q$. Moreover, if $O \leq Q \leq R$, then $R + Q \geq O$, and so $R - Q$ commutes with the unique nonnegative square root of $R + Q$. Since $T^* S T \in \mathcal{B}^+[\mathcal{H}]$ for every T in $\mathcal{B}[\mathcal{H}]$ and every S in $\mathcal{B}^+[\mathcal{H}]$, we get $R^2 - Q^2 = (R + Q)^{\frac{1}{2}} (R - Q)(R + Q)^{\frac{1}{2}} \geq O$, which completes the proof. However, in general (i.e., if Q and R do not commute),

$$Q \leq R \quad \text{does not imply} \quad Q^2 \leq R^2.$$

For instance, if $Q = \left(\begin{smallmatrix} 1 & 0 \\ 0 & 0 \end{smallmatrix}\right)$ and $R = \left(\begin{smallmatrix} 2 & 1 \\ 1 & 1 \end{smallmatrix}\right)$ in $\mathcal{B}^+[\mathbb{C}^2]$, then $R^2 - Q^2 = \left(\begin{smallmatrix} 4 & 3 \\ 3 & 2 \end{smallmatrix}\right)$ is not nonnegative. But the converse always holds. The purpose of the next problem is to prove the converse by using the previous problem.

Problem 8.12. Let A, B, Q and R be operators on a Hilbert space $\mathcal{H}$. Apply the Bounded Inverse Theorem to prove the following assertion.

(a) If A and B are self-adjoint, at least one of them being strictly positive, then $T = A + iB$ in $\mathcal{B}[\mathcal{H}]$ is invertible (i.e., $T \in \mathcal{G}[\mathcal{H}]$).

Now use (a) with Problem 8.11 and the Square Root Theorem to conclude:

(b) If Q and R are nonnegative operators, then

$$Q \leq R \quad \text{implies} \quad Q^{\frac{1}{2}} \leq R^{\frac{1}{2}}.$$

Another major result in spectral theory is the Fuglede Theorem. For a proof of it see, for instance, [47, p. 19].

Fuglede Theorem. *Let* $N = \int \lambda \, dP_\lambda$ *be the spectral decomposition of a normal operator* N. *If an operator* L *commutes with* N, *then* L *commutes with* $P(\Lambda)$ *for every* Λ *in* $\Sigma_{\sigma(N)}$.

Apply the Spectral Theorem and the Fuglede Theorem to prove the following assertion.

Problem 8.13. Every nonscalar normal operator (on a complex Hilbert space of dimension greater than 1) has a nontrivial hyperinvariant subspace that reduces every operator that commutes with it. In particular, every operator that commutes with a nonscalar normal operator is reducible.

Problem 8.14. The operators in this problem act on a complex Hilbert space of dimension greater than 1. Prove the following propositions.

(a) Every quasinormal operator has a nontrivial invariant subspace.

(b) Every isometry has a nontrivial invariant subspace.

Remarks: Problem 8.14 is easy; very easy. However if we go just one step ahead, from quasinormal to subnormal, we find a large gap but a bridgeable gap. *Every subnormal operator has a nontrivial invariant subspace.* This is a deep result that brought on a new technique for approaching invariant subspace problems. The original proof in [6] is not elementary at all but an easier and shorter proof can be found in [53] (also see [57, p. 97]). Another step ahead, from subnormal to hyponormal, lead us to an even larger gap, hitherto a bridgeless gap. In fact, it is still unknown whether every hyponormal operator has a nontrivial invariant subspace.

A rather useful corollary of the Fuglede Theorem reads as follows.

Fuglede–Putnam Theorem. *Let $\mathcal{H}$ and $\mathcal{K}$ be Hilbert spaces. If N_1 and N_2 are normal operators on $\mathcal{H}$ and $\mathcal{K}$, respectively, and if X in $\mathcal{B}[\mathcal{H}, \mathcal{K}]$ intertwines N_1 to N_2, then X intertwines N_1^* to N_2^* (i.e., if $XN_1 = N_2X$, then $XN_1^* = N_2^*X$).*

Problem 8.15. Prove the Fuglede–Putnam Theorem.

Problem 8.16. Let T be an operator acting on a complex Hilbert space. Show that the following assertions are pairwise equivalent.

(a) T is reducible.

(b) T commutes with a nontrivial orthogonal projection.

(c) T commutes with a nonscalar normal operator.

(d) There exists a nonscalar operator that commutes with T and T^*.

Normal operators (on a complex Hilbert space of dimension greater than 1) have a nontrivial invariant subspace (cf. Problem 8.13). Also compact operators (on a complex Banach space of dimension greater than 1) have a nontrivial invariant subspace, as we shall see in Chapter 12. This prompts the most celebrated open question in operator theory, namely, the *invariant subspace problem*: Does every operator (on an infinite-dimensional complex separable Hilbert space) have a nontrivial invariant subspace? All the above attributes are indispensable. On a finite-dimensional normed space every operator is compact, and so every operator on a complex finite-dimensional normed space of dimension greater than 1 has a nontrivial invariant subspace. On a nonseparable normed space every operator has a nontrivial invariant subspace. In fact, if T is an operator on a nonseparable normed space $\mathcal{X}$, then the span of the orbit of any nonzero vector x in $\mathcal{X}$ under T, $\bigvee\{T^n x\}_{n\geq 0}$, is a nonzero invariant subspace for T (as we saw in Chapter 1). Since $\{T^n x\}_{n\geq 0}$ is a countable set, it follows that $\bigvee\{T^n x\}_{n\geq 0} \neq \mathcal{X}$ (a normed space is spanned by a countable set if and only if it is separable — see e.g., [32, p. 211]). Hence $\bigvee\{T^n x\}_{n\geq 0}$ is a nontrivial invariant subspace for T. The invariant subspace problem trivially has a negative answer in a real space. For instance, the operator $\left(\begin{smallmatrix} 0 & 1 \\ -1 & 0 \end{smallmatrix}\right)$ on $\mathbb{R}^2$ has no nontrivial invariant subspace (when acting on the Euclidean *real* space but, of course, it has a nontrivial invariant subspace when acting on the *complex* space $\mathbb{C}^2$). The problem also has a negative answer if we replace Hilbert space with Banach space. This (the invariant subspace problem in a Banach space) remained as an open question for a long period, but a Banach space operator without a nontrivial invariant subspace was constructed in [48] and [17]. However, for operators on an infinite-dimensional complex separable Hilbert space, the invariant subspace problem remains a recalcitrant open question.

Solutions

Solution 8.1. Let T be an operator on a complex Hilbert space $\mathcal{H}$ and take an arbitrary $\lambda \in \mathbb{C}$. Put $S = (\lambda I - T)$ so that $S^* = (\bar{\lambda} I - T^*)$. Since $S \in \mathcal{G}[\mathcal{H}]$ if and only if $S^* \in \mathcal{G}[\mathcal{H}]$, it follows that

$$\rho(T) = \rho(T^*)^*,$$

and hence

$$\sigma(T) = \mathbb{C}\backslash\rho(T) = \mathbb{C}\backslash\rho(T^*)^* = \left(\mathbb{C}\backslash\rho(T^*)\right)^* = \sigma(T^*)^*.$$

If λ lies in $\sigma_C(T)$, then $\mathcal{N}(S) = \{0\}$ and $\mathcal{R}(S)^- = \mathcal{H}$ which implies, by Problem 2.1, that $\mathcal{R}(S^*)^- = \mathcal{H}$ and $\mathcal{N}(S^*) = \{0\}$. Moreover, $\mathcal{R}(S) \neq \mathcal{H}$ so that $\mathcal{R}(S)$ is not closed, and so $\mathcal{R}(S^*)$ is not closed as well. Therefore, $\mathcal{R}(S^*)^- \neq \mathcal{H}$. Then $\bar{\lambda}$ lies in $\sigma_C(T^*)$. Outcome: $\sigma_C(T) \subseteq \sigma_C(T^*)^*$. Since $T^{**} = T$, the previous inclusion implies that $\sigma_C(T^*) \subseteq \sigma_C(T)^*$ and, consequently, $\sigma_C(T^*)^* \subseteq \sigma_C(T)^{**} = \sigma_C(T)$. Thus

$$\sigma_C(T) = \sigma_C(T^*)^*.$$

Since $\mathcal{N}(S^*) = \{0\}$ if and only if $\mathcal{R}(S)^- = \mathcal{H}$ (Problem 2.1(b)), we get

$$\mathcal{R}(S)^- \neq \mathcal{H} \iff \mathcal{N}(S^*) \neq \{0\} \iff \bar{\lambda} \in \sigma_P(T^*) \iff \lambda \in \sigma_P(T^*)^*.$$

If λ lies in $\sigma_R(T)$, then $\mathcal{R}(S)^- \neq \mathcal{H}$ and $\lambda \notin \sigma_P(T)$ (recall that the residual spectrum is disjoint with the point spectrum) so that $\lambda \in \sigma_P(T^*)^* \backslash \sigma_P(T)$. Conversely, if λ lies in $\sigma_P(T^*)^* \backslash \sigma_P(T)$, then $\mathcal{R}(S)^- \neq \mathcal{H}$ and $\mathcal{N}(S) \neq \{0\}$ so that $\lambda \in \sigma_R(T)$. Therefore,

$$\sigma_R(T) = \sigma_P(T^*)^* \backslash \sigma_P(T).$$

Solution 8.2. Let T be an operator on a complex normed space $\mathcal{X}$.

(a) Take an arbitrary complex number λ. According to Problem 1.2(a), the subspace $\mathcal{N}(\lambda I - T)$ is invariant for every operator L that commutes with $\lambda I - T$, and hence for every L that commutes with T. That is, $\mathcal{N}(\lambda I - T)$ is a hyperinvariant subspace for T.

(b) An operator T is nonscalar if and only if $\mathcal{N}(\lambda I - T) \neq \mathcal{X}$ for every λ in $\mathbb{C}$. A complex number λ is an eigenvalue of T (i.e., $\lambda \in \sigma_P(T)$) if and only if $\mathcal{N}(\lambda I - T) \neq \{0\}$. If λ is an eigenvalue of T, then the subspace $\mathcal{N}(\lambda I - T)$ is an eigenspace of T. Therefore, if T is nonscalar and if λ is an eigenvalue of T, then the eigenspace $\mathcal{N}(\lambda I - T)$ is nontrivial, and hyperinvariant for T according to assertion (a).

(c) If $\mathcal{X}$ has nontrivial subspaces (equivalently, if dim $\mathcal{X} > 1$) and if T has no nontrivial invariant subspace, then T is nonscalar (reason: every nontrivial subspace of $\mathcal{X}$ is invariant for any scalar operator), and so $\sigma_P(T) = \varnothing$ according to assertion (b).

(d) Now consider an operator T acting on a complex Hilbert space $\mathcal{H}$ and suppose $\sigma_P(T) \cup \sigma_R(T) \neq \varnothing$. Since $\sigma_R(T) = \sigma_P(T^*)^* \backslash \sigma_P(T)$, this is equivalent to $\sigma_P(T) \cup \sigma_P(T^*) \neq \varnothing$ (see Problem 8.1). Therefore, either $\sigma_P(T) \neq \varnothing$ or $\sigma_P(T^*) \neq \varnothing$ (or both). Moreover, suppose T is nonscalar (or, equivalently, T^* is nonscalar). If $\sigma_P(T) \neq \varnothing$, then T has a nontrivial hyperinvariant subspace by assertion (b). Similarly, if $\sigma_P(T^*) \neq \varnothing$, then T^* has a nontrivial hyperinvariant subspace, whose orthogonal complement is a nontrivial hyperinvariant subspace for T (Problem 4.3). Summing up: if $\sigma_P(T) \cup \sigma_R(T) \neq \varnothing$ and T is nonscalar, then T has a nontrivial hyperinvariant subspace.

(e) If dim $\mathcal{H} > 1$ and T has no nontrivial invariant subspace, then T is nonscalar (see proof of (c) above), and hence $\sigma_P(T) \cup \sigma_R(T) = \varnothing$ by assertion (d); that is, $\sigma(T) = \sigma_C(T)$.

Solution 8.3. Take any $T \in \mathcal{B}[\mathcal{H}]$ and any $\lambda \in \mathbb{C}$. Note that

$$(\lambda I - T)^*(\lambda I - T) - (\lambda I - T)(\lambda I - T)^* = T^*T - TT^*.$$

Thus $(\lambda I - T)$ is hyponormal if and only if T is hyponormal. Suppose H is hyponormal so that $(\lambda I - H)$ is hyponormal. Hence (cf. Problem 7.3)

$$\|(\bar{\lambda}I - H^*)x\| \le \|(\lambda I - H)x\| \quad \text{for every } x \in \mathcal{H} \text{ and every } \lambda \in \mathbb{C}.$$

If $\lambda \in \sigma_P(H)$, then $\mathcal{N}(\lambda I - H) \ne \{0\}$ so that $\mathcal{N}(\bar{\lambda}I - H^*) \ne \{0\}$ by the above inequality and so $\bar{\lambda} \in \sigma_P(H^*)$. Thus $\sigma_P(H) \subseteq \sigma_P(H^*)^*$. Therefore,

$$\sigma_P(H)^* \subseteq \sigma_P(H^*) \quad \text{so that} \quad \sigma_R(H^*) = \sigma_P(H)^* \backslash \sigma_P(H^*) = \varnothing$$

(cf. Problem 8.1), which proves (a). Since N is normal if and only if it is both hyponormal and cohyponormal, the above argument also proves (b):

$$\sigma_P(N)^* = \sigma_P(N^*) \quad \text{so that} \quad \sigma_R(N) = \varnothing.$$

Now let U be a unitary operator (that is, a normal isometry). Since U is an isometry, $\big|\,|\lambda| - 1\big|\,\|x\| = \big|\,\|\lambda x\| - \|Ux\|\,\big| \le \|(\lambda I - U)x\|$ for every x in $\mathcal{H}$. If $|\lambda| \ne 1$, then $(\lambda I - U)$ is bounded below so that $\lambda \in \rho(U) \cup \sigma_R(U) = \rho(U)$ since $\sigma_R(U) = \varnothing$ by item (b). Thus $\lambda \notin \rho(U)$ implies $|\lambda| = 1$, proving (c):

$$\sigma(U) \subseteq \Gamma.$$

Next let A be a self-adjoint operator (i.e., $A^* = A$). Since $\langle x\,;Ax\rangle \in \mathbb{R}$, it follows that $\langle \beta x\,;(\alpha I - A)x\rangle \in \mathbb{R}$, and hence $\text{Re}\,\langle i\beta x\,;(\alpha I - A)x\rangle = 0$, for every $\alpha, \beta \in \mathbb{R}$ and every $x \in \mathcal{H}$. Therefore, with $\lambda = \alpha + i\beta$,

$$\begin{aligned}
\|(\lambda I - A)x\|^2 &= \|i\beta x + (\alpha I - A)x\|^2 \\
&= |\beta|^2\|x\|^2 + 2\,\text{Re}\,\langle i\beta x\,;(\alpha I - A)x\rangle + \|(\alpha I - A)x\|^2 \\
&= |\beta|^2\|x\|^2 + \|(\alpha I - A)x\|^2 \ge |\beta|^2\|x\|^2 = |\,\text{Im}\,\lambda|^2\|x\|^2
\end{aligned}$$

for every $x \in \mathcal{H}$ and every $\lambda \in \mathbb{C}$. If λ is not real, then $(\lambda I - A)$ is bounded below, and so $\lambda \in \rho(A) \cup \sigma_R(A) = \rho(A)$ because $\sigma_R(A) = \varnothing$ by item (b) once A is normal. Thus $\lambda \notin \rho(A)$ implies $\lambda \in \mathbb{R}$. Since $\sigma(A)$ is bounded, this shows that (d) holds:

$$\sigma(A) \subset \mathbb{R}.$$

If $Q \ge O$ and $\lambda \in \sigma(Q)$, then λ is real according to item (d) because Q is self-adjoint, and hence $\|(\lambda I - Q)x\|^2 = |\lambda|^2\|x\|^2 - 2\lambda\langle Qx\,;x\rangle + \|Qx\|^2$ for each x in $\mathcal{H}$. If $\lambda < 0$, then $\|(\lambda I - Q)x\|^2 \ge |\lambda|^2\|x\|^2$ for every $x \in \mathcal{H}$ (since Q is nonnegative), and so $(\lambda I - Q)$ is bounded below. Applying the same argument of the previous item we get (e):

$$\sigma(Q) \subset [0, \infty).$$

If $R \succ O$, then $O \le R \in \mathcal{G}[\mathcal{H}]$, and hence $0 \in \rho(R)$ and $\sigma(R) \subset [0, \infty)$ by item (e) so that $\sigma(R) \subset (0, \infty)$. But $\sigma(R)$ is closed. Thus (f) holds:

$$\sigma(R) \subset [\alpha, \infty) \quad \text{for some} \quad \alpha > 0.$$

Finally, let P be a nontrivial projection in $\mathcal{B}[\mathcal{H}]$. Since $O \ne P = P^2 \ne I$, it follows that $\{0\} \ne \mathcal{R}(P) = \mathcal{N}(I - P)$ and $\{0\} \ne \mathcal{R}(I - P) = \mathcal{N}(P)$, and so $\{0, 1\} \subseteq \sigma_P(T)$. If λ is any complex number such that $0 \ne \lambda \ne 1$, then

$$(\lambda I - P)\left(\tfrac{1}{\lambda}I + \tfrac{1}{\lambda(\lambda-1)}P\right) = I = \left(\tfrac{1}{\lambda}I + \tfrac{1}{\lambda(\lambda-1)}P\right)(\lambda I - P)$$

so that $(\lambda I - P)$ is invertible (i.e., $(\lambda I - P) \in \mathcal{G}[\mathcal{H}]$), and hence $\lambda \in \rho(T)$. Thus $\sigma(P) \subseteq \{0, 1\}$, which concludes the proof of (g):

$$\sigma(P) = \sigma_P(P) = \{0, 1\}.$$

Solution 8.4. Recall from Problem 8.1 that $\sigma_R(T) = \sigma_P(T^*)^* \backslash \sigma_P(T)$. If λ lies in $\sigma_R(T)$, then there exists a nonzero x in $\mathcal{H}$ such that $T^*x = \bar{\lambda}x$ and $Tx \neq \lambda x$, which implies that

$$0 < \|Tx - \lambda x\|^2 = \|Tx\|^2 - 2\operatorname{Re}\langle \bar{\lambda}x ; T^*x \rangle + |\lambda|^2\|x\|^2 = \|Tx\|^2 - |\lambda|^2\|x\|^2.$$

Therefore, $|\lambda| < \frac{\|Tx\|}{\|x\|} \leq \|T\|$. That is, $\sigma_R(T) \subseteq \{\lambda \in \mathbb{C}: |\lambda| < \|T\|\}$.

Solution 8.5. Let $T_+ = S_+D$ be a unilateral weighted shift, where S_+ is a canonical unilateral shift on ℓ_+^2 and $D = \operatorname{diag}(\{\alpha_k\}_{k=0}^\infty)$ in $\mathcal{B}[\ell_+^2]$. Take an arbitrary positive integer n. It is easy to show by induction that, for each $x = (\xi_0, \xi_1, \xi_2, \ldots)$ in ℓ_+^2,

$$T_+^n x = \left(0, \ldots, 0, \ \textstyle\prod_{j=0}^{n-1}\alpha_j\xi_0, \ \prod_{j=1}^{n}\alpha_j\xi_1, \ \prod_{j=2}^{n+1}\alpha_j\xi_2, \ \ldots, \ \prod_{j=k}^{n+k-1}\alpha_j\xi_k, \ \ldots\right),$$

where the first n entries (from position 0 to position $n-1$) are all zero and the entry at the position $n+k$ is $\prod_{j=k}^{n+k-1}\alpha_j\xi_k$. Therefore, $T_+^n = S_+^n D_n$ with $D_n = \operatorname{diag}(\{\prod_{j=k}^{n+k-1}\alpha_j\}_{k=0}^\infty)$ so that $\|T_+^n\| = \|D_n\| = \sup_k |\prod_{j=k}^{n+k-1}\alpha_j|$ (use the same argument as for $n = 1$ in Solution 5.9(b)). The Gelfand–Beurling formula for the spectral radius then ensures that

$$r(T_+) = \lim_n \|T_+^n\|^{\frac{1}{n}} = \lim_n \left(\sup_k \left|\prod_{j=k}^{n+k-1}\alpha_j\right|\right)^{\frac{1}{n}} = \lim_n \sup_k \left(\prod_{j=k}^{n+k-1}|\alpha_j|\right)^{\frac{1}{n}}.$$

Solution 8.6. Since $r(T)^n = r(T^n) \leq \|T^n\|$ for each positive integer n, it follows that $\|T^n\| \to 0$ implies $r(T) < 1$. Now suppose $r(T) < 1$ and take an arbitrary α in $(r(T), 1)$. Since $r(T) = \lim_n \|T^n\|^{\frac{1}{n}}$ (Gelfand–Beurling formula), there is an integer $n_\alpha \geq 1$ such that $\|T^n\| \leq \alpha^n$ for every $n \geq n_\alpha$. Thus $\|T^n\| \leq \beta\alpha^n$ for every $n \geq 0$ with $\beta = \max_{0 \leq n \leq n_\alpha} \|T^n\|\alpha^{-n_\alpha}$, which clearly implies $\|T^n\| \to 0$. Therefore, considering the assertions (a)–(c) in the problem statement, we have proved that (a)$\Rightarrow$(b)$\Rightarrow$(c)$\Rightarrow$(a).

Solution 8.7. Let $\mathcal{X}$ be a complex Banach space and take a finite-rank operator $T \in \mathcal{B}_0[\mathcal{X}]$. If $0 \notin \sigma_P(T)$, then $\mathcal{N}(T) = \{0\}$. Thus T is injective. Since $\dim \mathcal{N}(T) = 0$, it follows that $\dim \mathcal{X} = \dim \mathcal{R}(T)$, which is finite, and so $\mathcal{R}(T) = \mathcal{X}$. Thus the injective T also is surjective so that $T \in \mathcal{G}[\mathcal{X}]$ by the Inverse Mapping Theorem, and hence $0 \in \rho(T)$. Conclusion: either

$0 \in \sigma_P(T) \subseteq \sigma(T)$ or $0 \in \rho(T)$. But the Fredholm alternative ensures that $\sigma(T)\backslash\{0\} = \sigma_P(T)\backslash\{0\}$ since $\mathcal{B}_0[\mathcal{X}] \subseteq \mathcal{B}_\infty[\mathcal{X}]$. Therefore,

$$\sigma(T) = \sigma_P(T).$$

Suppose the point spectrum $\sigma_P(T)$ is not finite. Since eigenvectors associated with distinct eigenvalues are linearly independent, there exists an infinite set of linearly independent eigenvectors in $\mathcal{R}(T)$ so that $\mathcal{R}(T)$ is infinite-dimensional, which is a contradiction. Thus $\sigma_P(T)$ must be finite.

Solution 8.8. Let T be an operator on a complex Hilbert space $\mathcal{H}$.

(a) Suppose $\sigma_P(T) \neq \varnothing$ (otherwise (a) is trivially satisfied) and take an arbitrary λ in $\sigma_P(T)$. Thus there exists a unit vector e in $\mathcal{H}$ ($e \in \mathcal{H}$ such that $\|e\| = 1$) for which $T^n e = \lambda^n e$, and hence $|\lambda|^n = |\langle T^n e ; e\rangle|$, for every positive integer n. If $T^n \xrightarrow{w} O$, then $\langle T^n e ; e \rangle \to 0$ so that $|\lambda| < 1$. Therefore, $T^n \xrightarrow{w} O$ implies $\sigma_P(T) \subseteq \Delta$.

(b) Suppose T is compact. If $\sigma_P(T) \subseteq \Delta$, then the Fredholm alternative ensures that $r(T) < 1$, and hence $T^n \xrightarrow{u} O$ by the Gelfand–Beurling formula for the spectral radius (Problem 8.6).

(c) Recall that uniform stability implies strong stability, which in turn implies weak stability for any operator. If T is compact, then weak stability implies uniform stability according to items (a) and (b), and so the concepts of weak, strong, and uniform stabilities coincide.

Solution 8.9. Let T be a weakly stable operator on a Hilbert space so that T^* is again weakly stable (Problem 3.3). Thus $\sigma_P(T) \cup \sigma_P(T^*) \subseteq \Delta$ (Problem 8.8(a)). Since $\sigma_R(T) = \sigma_P(T^*)^* \backslash \sigma_P(T)$ (Problem 8.1), we get $\sigma_P(T) \cup \sigma_R(T) \subseteq \Delta$. Hence, if $T^n \xrightarrow{w} O$ and $\sigma_C(T) \cap \Gamma = \varnothing$, then $r(T) < 1$ so that $T^n \xrightarrow{u} O$ (Problem 8.6). The converse is trivial: uniform stability implies strong stability, which in turn implies weak stability.

Solution 8.10. If T is a contraction, then the Nagy–Foiaş–Langer decomposition says that $T = C \oplus U$ on $\mathcal{H} = \mathcal{U}^\perp \oplus \mathcal{U}$, where C is a completely nonunitary contraction and U is unitary (see Problem 6.7). Recall that A and A_* stand for the strong limits of $\{T^{*n}T^n\}$ and $\{T^n T^{*n}\}$, and let A' and A'_* denote the strong limits of $\{C^{*n}C^n\}$ and $\{C^n C^{*n}\}$, respectively. Therefore, $A = A' \oplus I$ and $A_* = A'_* \oplus I$ on $\mathcal{H} = \mathcal{U}^\perp \oplus \mathcal{U}$. Since C is a completely nonunitary contraction, it follows by Problem 6.10 that C is weakly stable. If T is compact, then so is C (direct summands of a compact operator are again compact). Thus C is a weakly stable compact operator. But for a compact operator (on a complex Hilbert space) weak stability coincides with uniform stability (Problem 8.8) so that both C and C^* are uniformly stable, and hence strongly stable; that is, $A' = A'_* = O$ (Problem 6.2(a)). Then $A = A_* = O \oplus I$. Outcome: if T is compact, then $A = A_*$.

Solution 8.11. Recall that Γ denotes the unit circle about the origin of the complex plane. Half of this problem comes out with Problem 8.3. The Spectral Theorem ensures the other half. Indeed, if T is normal, then consider its spectral decomposition so that

$$T = \int_{\sigma(T)} \lambda \, dP_\lambda, \quad T^* = \int_{\sigma(T)} \bar\lambda \, dP_\lambda, \quad T^*T = \int_{\sigma(T)} |\lambda|^2 \, dP_\lambda = TT^*.$$

(a) If $\sigma(T) \subseteq \Gamma$, then $T^*T = TT^* = \int_{\sigma(T)} dP_\lambda = P(\sigma(T)) = I$, which means that T is unitary.

(b) If $\sigma(T) \subset \mathbb{R}$, then $T^* = T$; that is, T is self-adjoint.

(c) If $\sigma(T) \subset [0,\infty)$, then set $\beta = \inf \sigma(T) = \min \sigma(T)$ so that $\beta \geq 0$. Therefore, $T = \int_{\sigma(T)} \lambda \, dP_\lambda \geq \beta \int_{\sigma(T)} dP_\lambda = \beta P(\sigma(T)) = \beta I \geq O$. In other words, T is nonnegative.

(d) If $\sigma(T) \subset [\alpha,\infty)$ for some $\alpha > 0$, then $\beta = \min \sigma(T) \geq \alpha > 0$, and so $T \geq \beta I$ for some $\beta > 0$. Equivalently, T is strictly positive.

(e) If $\sigma(T) \subseteq \{0,1\}$, then $T = \int_{\{1\}} \lambda \, dP_\lambda = P(\{1\})$ if 1 lies in $\sigma(T)$, or $T = O$ if $\sigma(T) = \{0\}$, which are orthogonal projections.

Solution 8.12. $\mathcal{H}$ is a complex Hilbert space.

(a) Let A and B be self-adjoint operators on $\mathcal{H}$ and put $T = A + iB$ in $\mathcal{B}[\mathcal{H}]$ so that $\operatorname{Re}\langle Tx \, ; x \rangle = \langle Ax \, ; x \rangle$ and $\operatorname{Im}\langle Tx \, ; x \rangle = \langle Bx \, ; x \rangle$ for each vector x, and hence $|\langle Ax \, ; x \rangle|^2 + |\langle Bx \, ; x \rangle|^2 = |\langle Tx \, ; x \rangle|^2 \leq \|Tx\|^2 \|x\|^2$ for every x in $\mathcal{H}$. Thus T is bounded below if one of A or B is strictly positive. Similarly, T^* is bounded below if one of A or B is strictly positive. Then $\mathcal{R}(T) = \mathcal{R}(T)^-$ and $\mathcal{N}(T) = \mathcal{N}(T^*) = \{0\}$ by the Bounded Inverse Theorem, which implies that $\mathcal{R}(T) = \mathcal{N}(T^*)^\perp = \mathcal{H}$ (cf. Problem 2.1(b)). Therefore, T has a bounded inverse on its range $\mathcal{R}(T) = \mathcal{H}$ (again, Bounded Inverse Theorem) so that T lies in $\mathcal{G}[\mathcal{H}]$.

(b) Let Q and R be nonnegative operators on $\mathcal{H}$, take an arbitrary $\lambda > 0$, and put $A = \lambda^2 I + 2\lambda R + R^2 - Q^2$, $B = i(QR - RQ)$, $C = \lambda I + R + Q$ and $D = \lambda I + R - Q$ in $\mathcal{B}[\mathcal{H}]$. It is readily verified that these are all self-adjoint, C is strictly positive (thus invertible), and $DC = A + iB$. Observe that, if $R^2 - Q^2 \geq O$, then A is strictly positive. Therefore, according to (a), DC is invertible, and so is D (it is the product of two invertible operators, viz., $D = (DC)C^{-1}$). But if D is invertible, then $-\lambda \in \rho(R - Q)$, which holds for all $\lambda > 0$. This implies that $\sigma(R - Q) \subset \mathbb{C} \backslash (-\infty, 0)$. Since $R - Q$ is self-adjoint, it follows by Problem 8.11 that $R - Q \geq O$. Summing up:

$$\text{if } Q^2 \leq R^2, \text{ then } Q \leq R.$$

The Square Root Theorem says that the nonnegative square root of a nonnegative operator is unique. Therefore,

$$O \leq Q \leq R \quad \text{implies} \quad O \leq Q^{\frac{1}{2}} \leq R^{\frac{1}{2}}.$$

Solution 8.13. Let N and L be operators on a complex Hilbert space of dimension greater than 1. Suppose N is normal. If L commutes with N, then L commutes with each $P(\Lambda)$ (i.e., $LP(\Lambda) = P(\Lambda)L$) by the Fuglede Theorem, and hence each subspace $\mathcal{R}(P(\Lambda))$ reduces L (cf. Problem 4.6).

Outcome: $\{\mathcal{R}(P(\Lambda))\}_{\Lambda \in \Sigma_{\sigma(N)}}$ *is a family of reducing subspaces for every operator that commutes with a normal operator* $N = \int \lambda \, dP_\lambda$.

If $\sigma(N)$ has a single point, say $\sigma(N) = \{\mu\}$, then

$$N = \int_{\{\mu\}} \lambda \, dP_\lambda = \mu \int_{\{\mu\}} dP_\lambda = \mu I$$

so that N is a scalar operator (and every subspace of $\mathcal{H}$ reduces N). Thus, if N is nonscalar (so that $\dim \mathcal{H} > 1$), then $\sigma(N)$ has more than one point. Suppose N is nonscalar and take a pair of distinct points in $\sigma(N)$, say λ and μ. Let Δ_λ denote the nonempty open disc of radius $\frac{1}{2}|\lambda - \mu|$ centered at λ. Put $\Lambda_\lambda = \sigma(N) \cap \Delta_\lambda$ and $\Lambda'_\lambda = \sigma(N) \backslash \Delta_\lambda$ in $\Sigma_{\sigma(N)}$, which make a partition of $\sigma(N)$; that is, $\Lambda_\lambda \cap \Lambda'_\lambda = \varnothing$ and $\Lambda_\lambda \cup \Lambda'_\lambda = \sigma(N)$, and therefore $I = P(\sigma(N)) = P(\Lambda_\lambda \cup \Lambda'_\lambda) = P(\Lambda_\lambda) + P(\Lambda'_\lambda)$. Since both $\sigma(N) \cap \Delta_\lambda$ and $\sigma(N) \backslash \Delta_\lambda^-$ are nonempty relatively open subsets of $\sigma(N)$, we get $P(\Lambda_\lambda) \neq O$ and $P(\Lambda'_\lambda) \neq O$ (because $\Lambda_\lambda = \sigma(N) \cap \Delta_\lambda$ and $\sigma(N) \backslash \Delta_\lambda^- \subseteq \Lambda'_\lambda$). Hence $O \neq P(\Lambda_\lambda) = I - P(\Lambda'_\lambda) \neq I$ so that

$$\{0\} \neq \mathcal{R}(P(\Lambda_\lambda)) \neq \mathcal{H}.$$

Conclusion: $\mathcal{R}(P(\Lambda_\lambda))$ *is a nontrivial reducing subspace for every operator that commutes with a nonscalar normal operator* $N = \int \lambda \, dP_\lambda$.

Solution 8.14. Take an operator T on a complex Hilbert space of dimension greater than 1 and consider its self-commutator $D_T = T^*T - TT^*$.

(a) If T is quasinormal (i.e., $D_T T = O$), then either $D_T = O$ or $D_T \neq O$ and $D_T T = O$. In the former case it is normal, and we know from Problem 8.13 that normal operators have nontrivial invariant subspaces. In the latter case $\mathcal{N}(D_T)$ and $\mathcal{R}(T)^-$ are nontrivial invariant subspaces for T (Problem 1.3).

(b) Every isometry is quasinormal ($T^*T = I$ implies $(T^*T - TT^*)T = O$).

Solution 8.15. (a) Let $N = \int \lambda \, dP_\lambda$ be a normal operator in $\mathcal{B}[\mathcal{H}]$, let Λ be an arbitrary set in $\Sigma_{\sigma(N)}$, and take $L \in \mathcal{B}[\mathcal{H}]$.

Claim. $LN = NL \iff LP(\Lambda) = P(\Lambda)L \iff LN^* = N^*L$.

Proof. If $LN = NL$, then $LP(\Lambda) = P(\Lambda)L$ by the Fuglede Theorem so that

$$\langle LN^*x\,;y\rangle \;=\; \langle N^*x\,;L^*y\rangle \;=\; \int \overline{\lambda}\,d\langle P_\lambda x\,;L^*y\rangle$$

$$=\; \int \overline{\lambda}\,d\langle LP_\lambda x\,;y\rangle \;=\; \int \overline{\lambda}\,d\langle P_\lambda Lx\,;y\rangle \;=\; \langle N^*Lx\,;y\rangle$$

for every $x,y \in \mathcal{H}$, and hence $LN^* = N^*L$. Conversely, if $LN^* = N^*L$, then $NL^* = L^*N$, which implies $P(\Lambda)L^* = L^*P(\Lambda)$ (Fuglede Theorem again) so that $LP(\Lambda) = P(\Lambda)L$ (since $P(\Lambda)$ is self-adjoint), and therefore

$$\langle LNx\,;y\rangle \;=\; \langle Nx\,;L^*y\rangle \;=\; \int \lambda\,d\langle P_\lambda x\,;L^*y\rangle$$

$$=\; \int \lambda\,d\langle LP_\lambda x\,;y\rangle \;=\; \int \lambda\,d\langle P_\lambda Lx\,;y\rangle \;=\; \langle NLx\,;y\rangle$$

for every $x,y \in \mathcal{H}$ so that $LN = NL$. $\square$

(b) Now take $N_1 \in \mathcal{B}[\mathcal{H}]$, $N_2 \in \mathcal{B}[\mathcal{K}]$, and $X \in \mathcal{B}[\mathcal{H},\mathcal{K}]$. Put

$$N = N_1 \oplus N_2 = \begin{pmatrix} N_1 & O \\ O & N_2 \end{pmatrix} \quad \text{and} \quad L = \begin{pmatrix} O & O \\ X & O \end{pmatrix}$$

in $\mathcal{B}[\mathcal{H} \oplus \mathcal{K}]$. If N_1 and N_2 are normal operators, then it is clear that N is normal. If $XN_1 = N_2X$, then $LN = NL$ and so $LN^* = N^*L$ by the above claim. Hence $XN_1^* = N_2^*X$.

Solution 8.16. Take T in $\mathcal{B}[\mathcal{H}]$, where $\mathcal{H}$ is a complex Hilbert space, and consider the following assertions.

(a) T is reducible.

(b) T commutes with a nontrivial orthogonal projection.

(c) T commutes with a nonscalar normal operator.

(d) There exists a nonscalar operator that commutes with T and T^*.

We know from Problem 4.6 that (a) is equivalent to (b), which trivially implies (c) (every orthogonal projection is normal and nontrivial projections are nonscalar). The Fuglede Theorem (see Solution 8.13) ensures that (c) implies (b); and the Fuglede–Putnam Theorem ensures that (c) implies (d) (if $TN = NT$, then $TN^* = N^*T$ so that $NT^* = T^*N$). Moreover, if there exists a nonscalar operator L such that $LT = TL$ and $LT^* = T^*L$ (so that $L^*T = TL^*$), then $(L + L^*)T = T(L + L^*)$. Note that $L + L^*$ is always self-adjoint (thus normal). If $L + L^*$ is nonscalar, then T commutes with a nonscalar normal operator. If $L + L^*$ is scalar, then the nonscalar L is normal ($L + L^* = \alpha I \implies L^*L = LL^* = \alpha I - L^2$) and, again, T commutes with a nonscalar normal operator. Thus (d) implies (c).

9

Paranormal Operators

Let T be an operator on a Hilbert space $\mathcal{H}$ and consider its absolute value $|T| = (T^*T)^{\frac{1}{2}}$. Note that the self-commutator of T can be written as

$$D_T = T^*T - TT^* = |T|^2 - |T^*|^2.$$

Since $T^*(T^*T - TT^*)T = T^*T^*TT - T^*TT^*T$, it follows that

$$T^*D_TT = |T^2|^2 - |T|^4.$$

An operator T is *quasihyponormal* if $T^*D_TT \geq O$. Equivalently,

T is quasihyponormal if and only if $|T|^4 \leq |T^2|^2$.

We call an operator T *semi-quasihyponormal* if $|T^2| - |T|^2 \geq O$. That is,

T is semi-quasihyponormal if and only if $|T|^2 \leq |T^2|$.

Problem 9.1. Prove the following propositions.

(a) Every hyponormal operator is quasihyponormal.

(b) If a quasihyponormal operator has dense range, then it is hyponormal.

(c) If there exists a quasihyponormal operator without a nontrivial invariant subspace, then it is hyponormal.

Problem 9.2. Take any T in $\mathcal{B}[\mathcal{H}]$ and prove the following assertions.

(a) If T is quasinormal, then $|T|^2 = |T^2|$.

(b) If T is hyponormal and $|T|^2 = |T^2|$, then T is quasinormal.

(c) Every quasihyponormal operator is semi-quasihyponormal.

An operator T on a Hilbert space $\mathcal{H}$ is *paranormal* if

$$\|Tx\|^2 \le \|T^2 x\| \, \|x\| \quad \text{for every} \quad x \in \mathcal{H}.$$

Note that paranormal operators can be defined on normed spaces.

Problem 9.3. Prove the following propositions.

(a) Every semi-quasihyponormal operator is paranormal.

(b) Every paranormal operator is normaloid.

Problem 9.4. The square of a paranormal operator is again paranormal.

Here is a useful alternative definition of paranormal [1] (also see [49]).

Problem 9.5. Let $\mathcal{H}$ be a Hilbert space and take $T \in \mathcal{B}[\mathcal{H}]$.

(a) Show that T is paranormal if and only if

$$T^{2*}T^2 - 2\lambda T^*T + \lambda^2 I \ge O \quad \text{for all} \quad \lambda > 0.$$

Equivalently, T is paranormal if and only if

$$|T|^2 \le \tfrac{1}{2}\big(\lambda^{-1}|T^2|^2 + \lambda I\big) \quad \text{for all} \quad \lambda > 0.$$

(b) Show that, for every operator $T \in \mathcal{B}[\mathcal{H}]$,

$$|T^2| \le \tfrac{1}{2}\big(\lambda^{-1}|T^2|^2 + \lambda I\big) \quad \text{for all} \quad \lambda > 0.$$

Give another proof that every semi-quasihyponormal is paranormal.

Note: For any operator T in $\mathcal{B}[\mathcal{H}]$ the inequality in (a) holds for all $\lambda \le 0$. Hence it holds for all real λ if and only if T is paranormal.

Problem 9.6. Let T be an operator on a Hilbert space. If T is strictly positive, positive, nonnegative, self-adjoint, normal, quasinormal, subnormal, hyponormal, quasihyponormal, semi-quasihyponormal, paranormal or normaloid, then so is every operator unitarily equivalent to it. Check it.

Problem 9.7. Let T_+ be a unilateral weighted shift on ℓ_+^2 with a *nonnegative* weight sequence $\{\alpha_k\}_{k=0}^{\infty}$. Verify that $|T_+|^2 = \mathrm{diag}(\alpha_0^2, \alpha_1^2, \alpha_2^2, \alpha_3^2, \dots)$ and $|T_+^*|^2 = \mathrm{diag}(0, \alpha_0^2, \alpha_1^2, \alpha_2^2, \dots)$, and hence the self-commutator of T_+ is a diagonal operator on ℓ_+^2,

$$T_+^* T_+ - T_+ T_+^* = \mathrm{diag}(\{\delta_k\}_{k=0}^{\infty}),$$

whose diagonal entries are $\delta_0 = \alpha_0^2$ and $\delta_{k+1} = \alpha_{k+1}^2 - \alpha_k^2$ for every $k \geq 0$. Note that T_+ is never normal (unless it is null — that is, $T_+^* T_+ = T_+ T_+^*$ if and only if $\delta_k = 0$ for every $k \geq 0$ or, equivalently, if and only if $\alpha_k = 0$ for every $k \geq 0$). Now prove the following propositions.

(a) T_+ is quasinormal if and only if $\alpha_k(\alpha_{k+1}^2 - \alpha_k^2) = 0$ for every $k \geq 0$.

   Thus it is easy to show by induction that T_+ *is quasinormal if and only if the sequence* $\{\alpha_k\}_{k=0}^{\infty}$ *is either constant or eventually constant with a null initial segment* (i.e., if and only if $\alpha_k \neq 0$ implies $\alpha_{k+j} = \alpha_k$ for all $j \geq 0$). *Hence, if* T_+ *is injective* (i.e., if $\alpha_k \neq 0$ for every $k \geq 0$; see Problem 5.9), *then* T_+ *is quasinormal if and only if* $\{\alpha_k\}_{k=0}^{\infty}$ *is a constant positive sequence* (i.e., $T_+ = \alpha S_+$ *for some* $\alpha > 0$).

(b) T_+ is hyponormal if and only if $\alpha_{k+1} - \alpha_k \geq 0$ for every $k \geq 0$.

   A unilateral weighted shift (*with nonnegative weights*) *is hyponormal if and only if its weight sequence is increasing.*

(c) T_+ is quasihyponormal if and only if $\alpha_k^2(\alpha_{k+1}^2 - \alpha_k^2) \geq 0$ for every $k \geq 0$.

(d) T_+ is semi-quasihyponormal if and only if $\alpha_k(\alpha_{k+1} - \alpha_k) \geq 0$ for every $k \geq 0$.

(e) T_+ is paranormal if and only if $\alpha_k^2 \alpha_{k+1}^2 - 2\lambda\alpha_k^2 + \lambda^2 \geq 0$ for all $\lambda > 0$ for every $k \geq 0$.

Problem 9.8. Hyponormality, quasihyponormality, semi-quasihyponormality and paranormality coincide for *every* unilateral weighted shift.

   In light of the equivalence in Problem 9.8, it is worth noticing that there exist hyponormal unilateral weighted shifts that are not subnormal and subnormal unilateral weighted shifts that are not quasinormal. Examples are supplied by Problems 9.7(a,b), 7.12 and 7.13, all of which involve shifts whose weight sequences have at least one repeated weight. Problem 7.14 gives a subnormal shift with distinct weights so that it is not quasinormal. We can exhibit a hyponormal shift with distinct weights that is not subnormal as follows. It was shown in [51] that if $0 < \alpha_0 < \alpha_1 < \alpha_2$ are arbitrarily chosen, then there exists a subnormal unilateral weighted shift with these three initial weights, but the forth one ($\alpha_3 > \alpha_2$) cannot be arbitrarily chosen: *If* $T_+ = \text{shift}(\{\alpha_k\}_{k=0}^{\infty})$ *with* $0 < \alpha_0 < \alpha_1 < \alpha_2$ *is subnormal, then*

$$\alpha_3^2 \geq \alpha_2^2 + \frac{\alpha_0^2}{\alpha_2^2} \frac{(\alpha_2^2 - \alpha_1^2)^2}{\alpha_1^2 - \alpha_0^2}.$$

This supplies a necessary condition for subnormality which leads to a useful way of constructing hyponormal weighted shifts with distinct weights that are not subnormal. For instance, if we take the subnormal weighted shift of Problem 7.14 and replace just α_3 (recall that $\alpha_3^2 = \frac{4}{5} = \frac{3}{4} + \frac{1}{20}$) with any positive α such that $\frac{3}{4} < \alpha^2 < \frac{3}{4} + \frac{1}{36}$, then we get a hyponormal unilateral weighted shift with distinct weights that is not subnormal.

Problem 9.9. Exhibit a normaloid unilateral weighted shift that is not paranormal.

Thus paranormal and normaloid are related by proper inclusion:

$$\text{Paranormal} \subset \text{Normaloid}$$

(cf. Problem 9.3(b)). Note that a backward unilateral shift S_+^* on ℓ_+^2 (see the Example that follows Problem 7.10) is another instance of a normaloid operator that is not paranormal. It is easy to produce examples of normaloid operators that are not paranormal. The problem below provides a useful necessary condition for paranormality.

Problem 9.10. Prove the following assertions.

(a) If T is paranormal, then $\mathcal{N}(T) = \mathcal{N}(T^2)$.

(b) The orthogonal direct sum of a nonzero orthogonal projection and a nonzero nilpotent contraction is normaloid but not paranormal.

(c) $T = \begin{pmatrix} 1 & 0 & 0 \\ 0 & 0 & 0 \\ 0 & 1 & 0 \end{pmatrix}$ in $\mathcal{B}[\mathbb{C}^3]$ is normaloid but not paranormal.

Let $\{A_k\}_{k=0}^{\infty}$ be a bounded sequence of operators on a Hilbert space $\mathcal{H}$ (each A_k is an operator on $\mathcal{H}$ and $\sup_{k \geq 0} \|A_k\| < \infty$). Now consider the Hilbert space $\ell_+^2(\mathcal{H}) = \bigoplus_{k=0}^{\infty} \mathcal{H}$ (the direct sum of countably infinite copies of $\mathcal{H}$) and let the operator $T = T(\{A_k\}_{k=0}^{\infty})$ in $\mathcal{B}[\ell_+^2(\mathcal{H})]$ be defined by

$$Tx = 0 \oplus \bigoplus_{k=1}^{\infty} A_{k-1} x_{k-1} \quad \text{so that} \quad T^*x = \bigoplus_{k=0}^{\infty} A_k^* x_{k+1}$$

for each $x = \bigoplus_{k=0}^{\infty} x_k$ in $\ell_+^2(\mathcal{H})$, where 0 is the origin of $\mathcal{H}$. Thus

$$T^*Tx = \bigoplus_{k=0}^{\infty} A_k^* A_k x_k \quad \text{and} \quad TT^*x = 0 \oplus \bigoplus_{k=1}^{\infty} A_{k-1} A_{k-1}^* x_k$$

for every vector $x = \bigoplus_{k=0}^{\infty} x_k$ in $\ell_+^2(\mathcal{H})$ so that

$$|T|^2 = T^*T = \bigoplus_{k=0}^{\infty} A_k^* A_k \quad \text{and} \quad |T^*|^2 = TT^* = 0 \oplus \bigoplus_{k=1}^{\infty} A_{k-1} A_{k-1}^*$$

in $\mathcal{B}[\ell_+^2(\mathcal{H})]$. These operators on $\ell_+^2(\mathcal{H})$ can be identified with the following infinite matrices of operators on $\mathcal{H}$.

$$T = \begin{pmatrix} O & & & \\ A_0 & O & & \\ & A_1 & O & \\ & & A_2 & O \\ & & & & \ddots \end{pmatrix}, \quad T^* = \begin{pmatrix} O & A_0^* & & & \\ & O & A_1^* & & \\ & & O & A_2^* & \\ & & & O & \\ & & & & \ddots \end{pmatrix},$$

$$|T|^2 = \begin{pmatrix} |A_0|^2 & & & \\ & |A_1|^2 & & \\ & & |A_2|^2 & \\ & & & |A_3|^2 \\ & & & & \ddots \end{pmatrix}, \quad |T^*|^2 = \begin{pmatrix} O & & & \\ & |A_0^*|^2 & & \\ & & |A_1^*|^2 & \\ & & & |A_2^*|^2 \\ & & & & \ddots \end{pmatrix}.$$

Problem 9.11. Show that $\|T\| = \sup_{k \geq 0} \|A_k\|$ and prove the following assertions.

(a) T is quasinormal if and only if $(|A_{k+1}|^2 - |A_k^*|^2)A_k = O$ for every $k \geq 0$.

(b) T is hyponormal if and only if $|A_{k+1}|^2 - |A_k^*|^2 \geq O$ for every $k \geq 0$.

(c) T is quasihyponormal if and only if $A_k^*|A_{k+1}|^2 A_k - |A_k|^4 \geq O$ for every $k \geq 0$.

(d) T is semi-quasihyponormal if and only if $(A_k^*|A_{k+1}|^2 A_k)^{\frac{1}{2}} - |A_k|^2 \geq O$ for every $k \geq 0$.

(e) T is paranormal if and only if $A_k^*|A_{k+1}|^2 A_k - 2\lambda|A_k|^2 + \lambda^2 I \geq O$ for all $\lambda > 0$ for every $k \geq 0$.

Problem 9.12. Let $\{P_k\}_{k=0}^{\infty}$ be a sequence of orthogonal projections acting on a Hilbert space $\mathcal{H}$. Set $A_k = P_k$ for each integer $k \geq 0$ and consider the operator $T = T(\{P_k\}_{k=0}^{\infty}) \in \mathcal{B}[\ell_+^2(\mathcal{H})]$ of Problem 9.11. Show that

(a) T is quasinormal if and only if $\{P_k\}_{k=0}^{\infty}$ is an increasing sequence;

(b) if $P_{k_0+1} \leq P_{k_0}$ and $P_{k_0+1} \neq P_{k_0}$ for some $k_0 \geq 0$, then T is not paranormal.

Problem 9.13. Let $\{A_k\}_{k=0}^{\infty}$ be a sequence of *self-adjoint* operators on a Hilbert space $\mathcal{H}$ such that $A_0 = Q$ and $A_k = R$ for all $k \geq 1$. Consider the operator $T = T(Q, R) \in \mathcal{B}[\ell_+^2(\mathcal{H})]$ of Problem 9.11 and show that

(a) T is hyponormal if and only if $R^2 - Q^2 \geq O$,

(b) T is quasihyponormal if and only if $QR^2Q - Q^4 \geq O$,

(c) T is semi-quasihyponormal if and only if $(QR^2Q)^{\frac{1}{2}} - Q^2 \geq O$,

(d) T is paranormal if and only if $QR^2Q - 2\lambda Q^2 + \lambda^2 I \geq O$ for all $\lambda > 0$.

Weighted shifts comprise a most useful class of operators for providing examples and counterexamples. However, as we saw in Problem 9.8, they cannot be used to show that the inclusions

Hyponormal $\subset$ Quasihyponormal $\subset$ Semi-quasihyponormal $\subset$ Paranormal

are proper. We shall verify now that these are, in fact, proper inclusions.

Problem 9.14. Identify operators in $\mathcal{B}[\mathbb{C}^2]$ with their 2×2 matrices with respect to the canonical basis for the Hilbert space $\mathbb{C}^2$. Let Q and R be self-adjoint operators in $\mathcal{B}[\mathbb{C}^2]$ and consider the operator $T = T(Q, R)$ of Problem 9.13 in $\mathcal{B}[\ell_+^2(\mathbb{C}^2)]$. Set $Q = \left(\begin{smallmatrix} 1 & 0 \\ 0 & 0 \end{smallmatrix}\right)$ and $R = \left(\begin{smallmatrix} 1 & 1 \\ 1 & 1 \end{smallmatrix}\right)^{\frac{1}{2}}$, and show that

(a) T is quasihyponormal but not hyponormal.

Now put $Q = \frac{1}{\sqrt{2}} \left(\begin{smallmatrix} 0 & 1 \\ 1 & 1 \end{smallmatrix}\right)$ and $R = \left(\begin{smallmatrix} 2 & 1 \\ 1 & 1 \end{smallmatrix}\right)^{\frac{1}{2}}$, and show that

(b) T is semi-quasihyponormal but not quasihyponormal.

Finally put $Q = \frac{1}{2} \left(\begin{smallmatrix} 2 & 1 \\ 1 & 1 \end{smallmatrix}\right)$ and $R = \left(\begin{smallmatrix} 10 & -3 \\ -3 & 1 \end{smallmatrix}\right)^{\frac{1}{2}}$, and show that

(c) T is paranormal but not semi-quasihyponormal.

Problem 9.15. Consider the operator $T = T(Q, R) \in \mathcal{B}[\ell_+^2(\mathcal{H})]$ of Problem 9.11, now with $A_0 = A_1 = Q$ and $A_k = R$ for all $k \geq 2$, where Q and R are again self-adjoint operators on $\mathcal{H}$. First verify that

$$T \text{ is hyponormal} \iff R^2 - Q^2 \geq O,$$

$$T^2 \text{ is hyponormal} \iff R^4 - Q^4 \geq O.$$

Here is another example of a hyponormal operator that is not subnormal, and of a quasihyponormal operator that is not hyponormal. Set $\mathcal{H} = \mathbb{C}^2$ and take $Q = \left(\begin{smallmatrix} 1 & 0 \\ 0 & 0 \end{smallmatrix}\right)$ and $R = \left(\begin{smallmatrix} 2 & 1 \\ 1 & 1 \end{smallmatrix}\right)^{\frac{1}{2}}$ in $\mathcal{B}[\mathbb{C}^2]$. Show that

(a) T is hyponormal but not subnormal,

(b) T^2 is quasihyponormal but not hyponormal.

The square of a paranormal operator is again paranormal (Problem 9.4). This is a property shared by paranormal but not by hyponormal operators (the square of the hyponormal operator in Problem 9.15 is not hyponormal). However, there are many important properties shared by both classes. For instance, *the restriction of a paranormal operator to an invariant subspace is again paranormal* (compare with Problem 7.6). Indeed, if $\mathcal{M}$ is an invariant subspace for a paranormal operator T, then, for every u in $\mathcal{M}$,

$$\|T|_{\mathcal{M}} u\|^2 = \|Tu\|^2 \leq \|T^2 u\| \|u\| = \|(T|_{\mathcal{M}})^2 u\| \|u\|.$$

Problem 9.16. Let T be an invertible operator on a Hilbert space $\mathcal{H}$ (i.e., take any T in $\mathcal{G}[\mathcal{H}]$). Prove the following assertions.

(a) If T is hyponormal, then T^{-1} is hyponormal.

Hint: If Q and R are invertible operators such that $O \leq Q \leq R$, then $O \leq R^{-1} \leq Q^{-1}$ (see e.g., [32, p. 431]). Also recall that $T^{*-1} = T^{-1*}$.

(b) If T is paranormal, then T^{-1} is paranormal.

We close this chapter with an inequality for paranormal operators borrowed from [37] and some of its applications. Note that, although stated in a Hilbert space $\mathcal{H}$, the next two problems still hold in a Banach space.

Problem 9.17. If an operator T in $\mathcal{B}[\mathcal{H}]$ is paranormal, then

$$\frac{\|Tx\|^n}{\|x\|^n} \leq \frac{\|T^n x\|}{\|x\|}$$

for every nonzero vector x in $\mathcal{H}$ and every positive integer n.

Problem 9.18. Use the previous problem to give another proof that every paranormal operator is normaloid.

The above inequality supplies a simplified proof of an important result on hyponormal contractions [46] as stated in Problem 9.20 below (also see [37], [44] and [56]). Let T be a contraction and let A be the strong limit of $\{T^{*n}T^n\}$. Use the inequality of Problem 9.17 to prove the next proposition.

Problem 9.19. If T is a cohyponormal contraction, then A is a projection.

Now apply the above result together with the decomposition of Problem 6.9 to prove the following assertion.

Problem 9.20. Every completely nonunitary cohyponormal contraction is strongly stable.

Problem 9.21. If T is a quasinormal contraction, then both A and A_* are projections.

Solutions

Solution 9.1. Let T be an operator on a Hilbert space $\mathcal{H}$.

(a) Since $\langle T^* D_T T x\, ;\, x \rangle = \|D_T^{\frac{1}{2}} Tx\|^2$ for every $x \in \mathcal{H}$ whenever $D_T \geq O$, it follows that every hyponormal operator is quasihyponormal.

(b) Suppose T is quasihyponormal ($T^*D_T T \geq O$) and has dense range ($\mathcal{R}(T)^- = \mathcal{H}$). Take any x in $\mathcal{R}(T)$ so that $x = Tu$ for some u in $\mathcal{H}$. Thus $\langle D_T x ; x \rangle = \langle T^* D_T Tu ; u \rangle \geq 0$. Since $\mathcal{R}(T)$ is dense in $\mathcal{H}$, this extends by continuity from $\mathcal{R}(T)$ to $\mathcal{H}$. That is, if $\mathcal{R}(T)^- = \mathcal{H}$, then every y in $\mathcal{H}$ is the limit of a sequence $\{x_n\}$ of vectors in $\mathcal{R}(T)$ and

$$|\langle D_T x_n ; x_n \rangle - \langle D_T y ; y \rangle| = |\langle D_T(x_n - y) ; x_n \rangle + \langle D_T y ; (x_n - y) \rangle|$$
$$\leq \|D_T\| \|x_n - y\| \sup \|x_n\| + \|D_T y\| \|x_n - y\|$$

so that $\langle D_T y ; y \rangle = \lim \langle D_T x_n ; x_n \rangle \geq 0$. Thus $D_T \geq O$. Outcome: A quasihyponormal operator with dense range is hyponormal.

(c) If an operator has no nontrivial invariant subspace, then it is quasi-invertible (Problem 1.2(b)); that is, $\mathcal{R}(T)^- = \mathcal{H}$ and $\mathcal{N}(T) = \{0\}$. If, in addition, it is quasihyponormal, then it is hyponormal by (b).

Solution 9.2. Let T be a Hilbert space operator.

(a) The nonnegative square root of any nonnegative operator is unique. Therefore, $|T|^2 = |T^2|$ if and only if $|T|^4 = |T^2|^2$ or, equivalently, if and only if $T^* D_T T = O$. Thus $D_T T = O$ implies $|T|^2 = |T^2|$.

(b) Now suppose $D_T \geq O$ and consider the above argument. In this case we get $|T|^2 = |T^2|$ if and only if $(D_T^{\frac{1}{2}} T)^* (D_T^{\frac{1}{2}} T) = O$, which means $D_T^{\frac{1}{2}} T = O$. Thus $D_T \geq O$ and $|T|^2 = |T^2|$ implies $D_T T = O$.

(c) This is a straightforward application of Problem 8.12(b).

Solution 9.3. Let $\mathcal{H}$ be a Hilbert space and take an arbitrary x in $\mathcal{H}$.

(a) $\|Tx\|^2 = \langle Tx ; Tx \rangle = \langle T^* Tx ; x \rangle = \langle |T|^2 x ; x \rangle$ for every operator T in $\mathcal{B}[\mathcal{H}]$. If $|T|^2 \leq |T^2|$, then $\langle |T|^2 x ; x \rangle \leq \langle |T^2| x ; x \rangle \leq \||T^2| x\| \|x\|$ by the Schwarz inequality. But $\||T^2| x\| = \|T^2 x\|$ (cf. Problem 3.9(c)). Thus $|T|^2 \leq |T^2|$ implies $\|Tx\|^2 \leq \|T^2 x\| \|x\|$.

(b) If $T \in \mathcal{B}[\mathcal{H}]$ is paranormal, then $\|T T^{n-1} x\|^2 \leq \|T^2 T^{n-1} x\| \|T^{n-1} x\|$ so that $\|T^n x\|^2 \leq \|T^{n+1} x\| \|T^{n-1} x\| \leq \|T^{n+1}\| \|T^{n-1}\| \|x\|^2$ for each integer $n \geq 1$. Thus $\|T^n\|^2 \leq \|T^{n+1}\| \|T^{n-1}\|$ for every $n \geq 1$. Hence T is normaloid by Problems 7.10(b) and 7.9.

Solution 9.4. Take an arbitrary x in $\mathcal{H}$. If $T \in \mathcal{B}[\mathcal{H}]$ is paranormal, then

$$\|T^2 x\|^2 \leq \|T^3 x\| \|Tx\| \quad \text{and} \quad \|T^3 x\|^2 \leq \|T^4 x\| \|T^2 x\|.$$

These inequalities imply that $\|T^2 x\|^2 \|T^3 x\|^2 \leq \|T^4 x\| \|T^3 x\| \|T^2 x\| \|Tx\|$, and hence $\|T^2 x\|^3 \|T^3 x\| \leq \|T^4 x\| \|T^2 x\|^2 \|Tx\| \leq \|T^4 x\| \|T^3 x\| \|Tx\|^2$ so that (since T is paranormal) $\|T^2 x\|^3 \leq \|T^4 x\| \|Tx\|^2 \leq \|T^4 x\| \|T^2 x\| \|x\|$. Therefore,

$$\|T^2 x\|^2 \leq \|T^4 x\| \|x\|,$$

and the square of a paranormal operator is paranormal. (It can be shown by induction that *any positive power of a paranormal operator is paranormal*.)

Solution 9.5. Take an arbitrary operator $T \in \mathcal{B}[\mathcal{H}]$.

(a) If α and β are nonnegative numbers, then $0 \le 2\alpha\beta \le \lambda^{-1}\alpha^2 + \lambda\beta^2$ for all positive λ (reason: $0 \le (\lambda^{-\frac{1}{2}}\alpha - \lambda^{\frac{1}{2}}\beta)^2$). If α and β are both positive, then set $\lambda_0 = \alpha\beta^{-1} > 0$ so that $0 < 2\alpha\beta = \lambda_0^{-1}\alpha^2 + \lambda_0\beta^2$. If any of them is null, then $0 = 2\alpha\beta = \inf_{\lambda>0} \lambda^{-1}\alpha^2 + \lambda\beta^2$. Therefore,

$$\alpha\beta = \inf_{\lambda>0} \tfrac{1}{2}(\lambda^{-1}\alpha^2 + \lambda\beta^2) \quad \text{for every} \quad \alpha, \beta \ge 0.$$

Now take an arbitrary x in $\mathcal{H}$. Put $\alpha = \|T^2 x\|$ and $\beta = \|x\|$ so that

$$\|T^2 x\|\,\|x\| = \inf_{\lambda>0} \tfrac{1}{2}(\lambda^{-1}\|T^2 x\|^2 + \lambda\|x\|^2) = \inf_{\lambda>0} \tfrac{1}{2}\langle(\lambda^{-1}|T^2|^2 + \lambda I)x\,;x\rangle.$$

But $\langle|T|^2 x\,;x\rangle = \langle T^*Tx\,;x\rangle = \langle Tx\,;Tx\rangle = \|Tx\|^2$, and so T is paranormal if and only if $\langle|T|^2 x\,;x\rangle \le \|T^2 x\|\,\|x\|$; that is, if and only if

$$\langle|T|^2 x\,;x\rangle \le \inf_{\lambda>0} \tfrac{1}{2}\langle(\lambda^{-1}|T^2|^2 + \lambda I)x\,;x\rangle$$

or, equivalently, if and only if $|T|^2 \le \tfrac{1}{2}(\lambda^{-1}|T^2|^2 + \lambda I)$ for all $\lambda > 0$.

(b) If A is self-adjoint, then $O \le (\alpha A - \beta I)^2 = \alpha^2 A^2 - 2\alpha\beta A + \beta^2 I$ for every $\alpha, \beta \in \mathbb{R}$. Put $A = |T^2|$, $\alpha = \left(\frac{1}{2\lambda}\right)^{\frac{1}{2}}$ and $\beta = \left(\frac{\lambda}{2}\right)^{\frac{1}{2}}$ to get (b). That is, for every $T \in \mathcal{B}[\mathcal{H}]$, $|T^2| \le \tfrac{1}{2}(\lambda^{-1}|T^2|^2 + \lambda I)$ for all $\lambda > 0$.

Thus $|T|^2 \le |T^2|$ implies the inequality in (a): $|T|^2 \le \tfrac{1}{2}(\lambda^{-1}|T^2|^2 + \lambda I)$ for all $\lambda > 0$. Outcome: Every semi-quasihyponormal operator is paranormal.

Solution 9.6. Let $\mathcal{H}$ and $\mathcal{K}$ be unitarily equivalent Hilbert spaces. Take any operator T in $\mathcal{B}[\mathcal{H}]$ and any unitary transformation U in $\mathcal{G}[\mathcal{K}, \mathcal{H}]$ (U is invertible and $U^{-1} = U^*$). Set $S = U^*TU$ in $\mathcal{B}[\mathcal{K}]$ so that $S^* = U^*T^*U$. Since $S^*S = U^*T^*TU$, we get $|S| = U^*|T|U$ (by uniqueness of the nonnegative square root). Therefore, $|S| = S$ if and only if $|T| = T$ and, in this case, $|S| > O$ or $|S| \succ O$ if and only if $|T| > O$ or $|T| \succ O$, respectively. In other words, one of S or T is strictly positive, positive, or nonnegative if and only if the other is. Note that $S - S^* = U^*(T - T^*)U$, and hence S is self-adjoint if and only if T is. Moreover, the self-commutators are related by $D_S = U^*D_TU$, and so $D_S S = U^*D_TTU$ and $S^*D_S S = U^*T^*D_TTU$. Therefore, S is normal, quasinormal, hyponormal or quasihyponormal if and only if T is. Since $S^{2*}S^2 = U^*T^{2*}T^2U$, that is, since $|S^2| = U^*|T^2|U$, it follows that $O \le |T^2| - |T|^2$ if and only if $O \le |S^2| - |S|^2$. Thus one of S or T is semi-quasihyponormal if and only if the other is. Furthermore, the same argument with a little help from Problem 9.5 ensures that S is paranormal if and only if T is. Clearly, $S^n = U^*T^nU$ so that $\|S^n\| = \|T^n\|$ for every integer $n \ge 0$. Then (Problem 7.9) S is normaloid if and only if T

is normaloid. Finally, recall from Problem 7.11 that S is subnormal if and only if T is subnormal.

Solution 9.7. Let $T_+ = S_+D$ be a unilateral weighted shift on ℓ_+^2 with a nonnegative weight sequence $\{\alpha_k\}_{k=0}^\infty$ and compute

$$|T_+|^2 = T_+^* T_+ = D^* S_+^* S_+ D = D^* D = D^2 = \text{diag}(\{\alpha_k^2\}_{k=0}^\infty),$$

$$|T_+^*|^2 = T_+ T_+^* = S_+ DD^* S_+^* = S_+ D^2 S_+^* = \text{shift}(\{\alpha_k^2\}_{k=0}^\infty)S_+^*.$$

But

$$\text{diag}(\{\alpha_k^2\}_{k=0}^\infty) = \alpha_0 \oplus \text{diag}(\{\alpha_k^2\}_{k=1}^\infty),$$

$$\text{shift}(\{\alpha_k^2\}_{k=0}^\infty)S_+^* = 0 \oplus \text{diag}(\{\alpha_{k-1}^2\}_{k=1}^\infty).$$

Thus

$$T_+^* T_+ - T_+ T_+^* = \alpha_0 \oplus \text{diag}(\{\alpha_k^2 - \alpha_{k-1}^2\}_{k=1}^\infty) = \text{diag}(\{\delta_k\}_{k=0}^\infty),$$

with $\delta_0 = \alpha_0^2$ and $\delta_{k+1} = \alpha_{k+1}^2 - \alpha_k^2$ for every $k \geq 0$. Therefore,

$$(T_+^* T_+ - T_+ T_+^*)T_+ = \text{diag}(\{\delta_k\}_{k=0}^\infty)\text{shift}(\{\alpha_k\}_{k=0}^\infty) = \text{shift}(\{\alpha_k(\alpha_{k+1}^2 - \alpha_k^2)\}_{k=0}^\infty).$$

Hence

(a) $(T_+^* T_+ - T_+ T_+^*)T_+ = O$ if and only if $\alpha_k(\alpha_{k+1}^2 - \alpha_k^2) = 0$ for every $k \geq 0$,

(b) $T_+^* T_+ - T_+ T_+^* \geq O$ if and only if $\alpha_{k+1}^2 \geq \alpha_k^2$ for every $k \geq 0$.

Now compute

$$|T_+^2|^2 = T_+^{2*} T_+^2 = DS_+^* D^2 S_+ D = \text{diag}(\{\alpha_k^2 \alpha_{k+1}^2\}_{k=0}^\infty),$$

$$|T_+^2|^2 - |T_+|^4 = \text{diag}(\{\alpha_k^2 \alpha_{k+1}^2 - \alpha_k^4\}_{k=0}^\infty),$$

$$|T_+^2| - |T_+|^2 = \text{diag}(\{\alpha_k \alpha_{k+1} - \alpha_k^2\}_{k=0}^\infty).$$

Thus

(c) $|T_+^2|^2 - |T_+|^4 \geq O$ if and only if $\alpha_k^2(\alpha_{k+1}^2 - \alpha_k^2) \geq 0$ for every $k \geq 0$,

(d) $|T_+^2| - |T_+|^2 \geq O$ if and only if $\alpha_k(\alpha_{k+1} - \alpha_k) \geq 0$ for every $k \geq 0$.

Since

$$T_+^{2*} T_+^2 - 2\lambda T_+^* T_+ + \lambda^2 I = \text{diag}(\{\alpha_k^2 \alpha_{k+1}^2 - 2\lambda \alpha_k^2 + \lambda^2\}_{k=0}^\infty),$$

(e) $T_+^{2*} T_+^2 - 2\lambda T_+^* T_+ + \lambda^2 I \geq 0$ if and only if $\alpha_k^2 \alpha_{k+1}^2 - 2\lambda \alpha_k^2 + \lambda^2 \geq 0$ for every $k \geq 0$, for all $\lambda > 0$.

Solution 9.8. Consider the results (b)–(e) in Problem 9.7. The first part of the proof consists in showing that those results are equivalent. Indeed, (e)$\Rightarrow$(c) by completing the squares in (e); that is, $\alpha_k^2 \alpha_{k+1}^2 + (\alpha_k^2 - \lambda)^2 \geq \alpha_k^4$

for all $\lambda > 0$ implies $\alpha_k^4 - \alpha_k^2 \alpha_{k+1}^2 \leq \inf_{\lambda>0}(\alpha_k^2 - \lambda)^2 = 0$, for every $k \geq 0$. Moreover, (d)$\Rightarrow$(b) by contradiction: in fact, if (d) holds and (b) fails, then $\alpha_{k_0}(\alpha_{k_0+1} - \alpha_{k_0}) \geq 0$ and also $(\alpha_{k_0+1} - \alpha_{k_0}) < 0$ for some $k_0 \geq 0$ so that $\alpha_{k_0} = 0$, and hence $\alpha_{k_0+1} < 0$; which is a contradiction. Now recall that the implications (b)$\Rightarrow$(c)$\Rightarrow$(d)$\Rightarrow$(e) hold by Problems 9.1(a), 9.2(c) and 9.3(a). Thus, for unilateral weighted shifts with nonnegative weights, we get (b)$\Leftrightarrow$(c)$\Leftrightarrow$(d)$\Leftrightarrow$(e). According to Problems 5.10 and 9.6 this equivalence holds for *every* unilateral weighted shift (with any bounded complex weight sequences).

Solution 9.9. Let $T_+ = \text{shift}(\{\alpha_k\}_{k=0}^\infty)$ be a unilateral weighted shift on ℓ_+^2 with $\alpha_0 = 1$, $\alpha_1 = 0$, and $\alpha_k = 1$ for all $k \geq 2$. It is not hyponormal by Problem 9.7, and therefore (as a weighted shift) it is not paranormal by Problem 9.8. Since $\|T_+\| = 1$ and $\|T_+^n\| = \|T_+^2\| \neq 0$ for all $n \geq 2$, we get $\lim_n \|T_+^n\|^{\frac{1}{n}} = \|T_+\|$; that is, T_+ is normaloid ($r(T_+) = \|T_+\|$).

Solution 9.10. Let T be an operator on a Hilbert space $\mathcal{H}$.

(a) By definition, T is paranormal if $\|Tx\|^2 \leq \|T^2x\|\|x\|$ for every $x \in \mathcal{H}$. Thus, if T is paranormal, then $\mathcal{N}(T^2) \subseteq \mathcal{N}(T)$. But $\mathcal{N}(T) \subseteq \mathcal{N}(T^2)$ for every T. Therefore, $\mathcal{N}(T) = \mathcal{N}(T^2)$ whenever T is paranormal.

(b) Let $\mathcal{H}$ and $\mathcal{K}$ be Hilbert spaces. Take an orthogonal projection P on $\mathcal{H}$ and a nilpotent operator S on $\mathcal{K}$ such that $0 \neq \|S\| \leq \|P\|$. Set

$$T = P \oplus S \quad \text{in} \quad \mathcal{B}[\mathcal{H} \oplus \mathcal{K}].$$

Since $P^2 = P$, $S^2 = O$ and $\|S\| \leq \|P\| = 1$, we get $\|T^n\| = \|T\| = 1$ for all $n \geq 1$ so that T is normaloid. Indeed,

$$r(T) = \lim_n \|T^n\|^{\frac{1}{n}} = \|T\|.$$

Since $S \neq O$ and $S^2 = O$, it follows that $\mathcal{N}(S) \neq \mathcal{K} = \mathcal{N}(S^2)$, and hence the proper inclusion

$$\mathcal{N}(T) = \mathcal{N}(P) \oplus \mathcal{N}(S) \subset \mathcal{N}(P) \oplus \mathcal{N}(S^2) = \mathcal{N}(T^2),$$

which implies that T is not paranormal according to (a).

(c) $T = 1 \oplus \left(\begin{smallmatrix} 0 & 0 \\ 1 & 0 \end{smallmatrix}\right)$; an orthogonal projection $\oplus$ a nilpotent contraction.

Solution 9.11. Take an arbitrary vector $x = \bigoplus_{k=0}^\infty x_k$ in $\ell_+^2(\mathcal{H})$. Since $T^*Tx = \bigoplus_{k=0}^\infty A_k^*A_k x_k$, we get $\|T^*Tx\|^2 = \sum_{k=0}^\infty \|A_k^*A_k x_k\|^2$. Hence

$$\|T\|^4 = \|T^*T\|^2 = \sup_{\|x\|=1} \|T^*Tx\|^2 = \sup_{\|x\|=1} \sum_{k=0}^\infty \|A_k^*A_k x_k\|^2.$$

Now take an arbitrary unit vector e in $\mathcal{H}$. For each integer $j \geq 0$ put $e_{j,j} = e$ and $e_{j,k} = 0$ for every integer $k \geq 0$ such that $k \neq j$. Set $e_j = \bigoplus_{k=0}^{\infty} e_{j,k}$ in $\ell_+^2(\mathcal{H})$ (that is, set $e_j = (0, \ldots, 0, e, 0, \ldots)$ with the only nonzero entry at the jth position) so that $\|e_j\| = \|e\| = 1$ for every $j \geq 0$. Thus

$$\|T\|^4 \geq \sup_{\|e\|=1} \sum_{k=0}^{\infty} \|A_k^* A_k e_{j,k}\|^2 = \sup_{\|e\|=1} \|A_j^* A_j e\|^2 = \|A_j^* A_j\|^2 = \|A_j\|^4$$

for all $j \geq 0$, and therefore $\|T\|^4 \geq \sup_{j \geq 0} \|A_j\|^4$. On the other hand,

$$\|T\|^4 \leq \sup_{\|x\|=1} \sup_{j \geq 0} \|A_j^* A_j\|^2 \sum_{k=0}^{\infty} \|x_k\|^2 = \sup_{\|x\|=1} \sup_{j \geq 0} \|A_j^* A_j\|^2 \|x\|^2 = \sup_{j \geq 0} \|A_j\|^4.$$

Outcome:

$$\|T\| = \sup_{j \geq 0} \|A_j\|.$$

Consider the assertions (a)–(e) in the problem statement. To prove assertions (a) and (b), set $\Delta_k = |A_{k+1}|^2 - |A_k^*|^2$ in $\mathcal{B}[\mathcal{H}]$ for each $k \geq 0$ so that $D_T = |T|^2 - |T^*|^2 = |A_0|^2 \oplus \bigoplus_{k=1}^{\infty} \Delta_{k-1}$ and

$$D_T T = \begin{pmatrix} O \\ \Delta_0 A_0 & O \\ & \Delta_1 A_1 & O \\ & & \Delta_2 A_2 & O \\ & & & & \ddots \end{pmatrix}.$$

(a) $D_T T = O$ if and only if $\Delta_k A_k = O$ every $k \geq 0$.

(b) $D_T \geq O$ if and only if $\Delta_k \geq O$ for every $k \geq 0$.

Since $|T^2|^2 = T^{2*} T^2 = \bigoplus_{k=0}^{\infty} A_k^* |A_{k+1}|^2 A_k$ and $|T|^2 = T^* T = \bigoplus_{k=0}^{\infty} |A_k|^2$,

(c) $|T^2|^2 - |T|^4 \geq O$ if and only if $A_k^* |A_{k+1}|^2 A_k - |A_k|^4 \geq O$ for all $k \geq 0$,

(d) $|T^2| - |T|^2 \geq O$ if and only if $(A_k^* |A_{k+1}|^2 A_k)^{\frac{1}{2}} - |A_k|^2 \geq O$ for all $k \geq 0$,

(e) $T^{2*} T^2 - 2\lambda T^* T + \lambda^2 I \geq O$ for every positive number λ if and only if $A_k^* |A_{k+1}|^2 A_k - 2\lambda |A_k|^2 + \lambda^2 I \geq O$ for all $k \geq 0$ and every $\lambda > 0$.

Solution 9.12. If P is an orthogonal projection, then $|P|^2 = P^* P = P^2 = P$.

(a) Suppose each P_k is an orthogonal projection. According to the above identity $(|P_{k+1}|^2 - |P_k^*|^2) P_k = P_{k+1} P_k - P_k$. Thus T is quasinormal if and only if $P_{k+1} P_k = P_k$ for every $k \geq 0$ (Problem 9.11(a)), which means that $P_k \leq P_{k+1}$ for every $k \geq 0$ (Problem 2.9).

(b) $P_{k_0+1} \leq P_{k_0}$ if and only if $\mathcal{R}(P_{k_0+1}) \subseteq \mathcal{R}(P_{k_0})$. If $\mathcal{R}(P_{k_0}) = \mathcal{R}(P_{k_0+1})$, then $P_{k_0} \leq P_{k_0+1} \leq P_{k_0}$ so that $P_{k_0} = P_{k_0+1}$. Thus, if there exists an integer $k_0 \geq 0$ such that $P_{k_0+1} \leq P_{k_0}$ and $P_{k_0+1} \neq P_{k_0}$, then $\mathcal{R}(P_{k_0+1})$ is properly included in $\mathcal{R}(P_{k_0})$ and $P_{k_0} P_{k_0+1} P_{k_0} = P_{k_0+1}$ (cf. Problem 2.9). Consider the operator

$$Q = P_{k_0}^* |P_{k_0+1}|^2 P_{k_0} - 2|P_{k_0}|^2 + I = P_{k_0+1} - 2P_{k_0} + I.$$

Note that $Q \not\geq O$. Indeed, take any x in $\mathcal{R}(P_{k_0}) \backslash \mathcal{R}(P_{k_0+1})$ so that

$$\langle Qx \, ; x \rangle = \langle P_{k_0+1}x \, ; x \rangle - 2\langle P_{k_0}x \, ; x \rangle + \langle x \, ; x \rangle = \|P_{k_0+1}x\|^2 - \|x\|^2 < 0$$

(because $P_{k_0}x = x$ and $\|P_{k_0+1}x\| < \|x\|$). Thus T is not paranormal: set $\lambda = 1$ in Problem 9.11(e).

Solution 9.13. Each operator A_k is self-adjoint so that $|A_k|^2 = A_k^2 \geq O$ and $A_k^* |A_{k+1}|^2 A_k = A_k A_{k+1}^2 A_k \geq O$ for every $k \geq 0$. Thus assertions (a), (b) and (c) follow from assertions (b), (c) and (e) in Problem 9.11, respectively (recall that the nonnegative square root of a nonnegative operator is unique). R^2 is nonnegative and nonnegative operators are trivially paranormal so that $R^4 - 2\lambda R^2 + \lambda^2 I \geq O$ for all $\lambda > 0$ by Problem 9.5. Therefore, assertion (d) follows from assertion (e) in Problem 9.11.

Solution 9.14. Consider the conditions (a)–(d) in Problem 9.13.

(a) T is quasihyponormal because $QR^2Q = Q^4$ but $R^2 - Q^2 = \left(\begin{smallmatrix} 0 & 1 \\ 1 & 1 \end{smallmatrix}\right)$ is not nonnegative so that T is not hyponormal.

(b) Compute $(QR^2Q)^{\frac{1}{2}} = \frac{1}{2}\left(\begin{smallmatrix} 1 & 1 \\ 1 & 3 \end{smallmatrix}\right)$ so that $(QR^2Q)^{\frac{1}{2}} - Q^2 = \frac{1}{2}\left(\begin{smallmatrix} 0 & 0 \\ 0 & 1 \end{smallmatrix}\right) \geq O$, and hence T is semi-quasihyponormal. But T is not quasihyponormal because $QR^2Q - Q^4 = \frac{1}{4}\left(\begin{smallmatrix} 0 & 1 \\ 1 & 5 \end{smallmatrix}\right)$ is not nonnegative.

(c) Compute $(QR^2Q)^{\frac{1}{2}} = \frac{1}{2}\left(\begin{smallmatrix} 5 & 2 \\ 2 & 1 \end{smallmatrix}\right)$ so that $(QR^2Q)^{\frac{1}{2}} - Q^2 = \frac{1}{4}\left(\begin{smallmatrix} 5 & 1 \\ 1 & 0 \end{smallmatrix}\right)$ is not nonnegative. Thus T is not semi-quasihyponormal. Also compute

$$X(\lambda) = QR^2Q - 2\lambda Q^2 + \lambda^2 I = \frac{1}{4}\begin{pmatrix} 4\lambda^2 - 10\lambda + 29 & 12 - 6\lambda \\ 12 - 6\lambda & 4\lambda^2 - 4\lambda + 5 \end{pmatrix}.$$

Since $4\lambda^2 - 10\lambda + 29 > 0$ and $4\lambda^2 - 4\lambda + 5 > 0$ for all $\lambda > 0$, it follows that $X(\lambda) \geq O$ for all $\lambda > 0$ if and only if $\det X(\lambda) \geq 0$ for all $\lambda > 0$. Now we can show that $\det X(\lambda) > 0$ for all $\lambda > 0$ as follows. Let p and f be real-valued functions defined on $(0, \infty)$ by $p(\lambda) = 4\lambda^2 - 14\lambda + 35$ and $f(\lambda) = \frac{22\lambda - 1}{4\lambda^2}$. Set $\lambda_1 = \frac{1}{11}$, $\lambda_2 = \frac{1}{4}(7 - \sqrt{30})$ and $\lambda_3 = \frac{7}{4}$ so that $\lambda_1 < \lambda_2 < \lambda_3$. Check that $f'(\lambda) = 0$ only at $\lambda = \lambda_1$ and $p'(\lambda) = 0$ only at $\lambda = \lambda_3$, and verify:

$$\max_{\lambda > 0} f(\lambda) = f(\lambda_1) = p(\lambda_2) \quad \text{and} \quad f(\lambda_2) < p(\lambda_3) = \min_{\lambda > 0} p(\lambda).$$

Thus $f(\lambda) < p(\lambda)$ for every $\lambda > 0$. But $\det X(\lambda) = \frac{\lambda^2}{4}\left(p(\lambda) - f(\lambda)\right)$. Hence $\det X(\lambda) > 0$ so that $X(\lambda) = QR^2Q - 2\lambda Q^2 + \lambda^2 I > O$ for all $\lambda > 0$, and therefore T is paranormal.

Solution 9.15. Q and R are self-adjoint so that $|Q|^2 = Q^2$ and $|R|^2 = R^2$. Thus (Problem 9.11(b)) T is hyponormal if and only if $R^2 - Q^2 \geq O$. Since

$$|T^2|^2 = T^{2*}T^2 = Q^4 \oplus QR^2Q \oplus R^4 \oplus R^4 \oplus \bigoplus_{k=4}^{\infty} R^4,$$

$$|T^{2*}|^2 = T^2 T^{2*} = O \oplus O \oplus Q^4 \oplus RQ^2R \oplus \bigoplus_{k=4}^{\infty} R^4,$$

it follows that

$$D_{T^2} = |T^2|^2 - |T^{2*}|^2 = Q^4 \oplus QR^2Q \oplus (R^4 - Q^4) \oplus R(R^2 - Q^2)R \oplus \bigoplus_{k=4}^{\infty} O.$$

Recall: Q^4 and QR^2Q are nonnegative operators because both Q and R are self-adjoint. Problem 8.12 says that $R^4 - Q^4 \geq O$ implies $R^2 - Q^2 \geq O$, which clearly implies $R(R^2 - Q^2)R \geq O$. Therefore, $D_{T^2} \geq O$ if and only if $R^4 - Q^4 \geq O$. Hence T^2 is hyponormal if and only if $R^4 - Q^4 \geq O$.

(a) $R^2 - Q^2 = \begin{pmatrix} 1 & 1 \\ 1 & 1 \end{pmatrix} \geq O$ but $R^4 - Q^4 = \begin{pmatrix} 4 & 3 \\ 3 & 2 \end{pmatrix}$ is not nonnegative. Thus T is hyponormal but T^2 is not. Then T^2 is not subnormal (subnormal operators are hyponormal), and hence T is not subnormal (the square of a subnormal operator is subnormal).

(b) $Q^2(R^4 - Q^4)Q^2 = \begin{pmatrix} 4 & 0 \\ 0 & 0 \end{pmatrix} \geq O$ and $QR^2(R^2 - Q^2)R^2Q \geq O$ because $R^2 - Q^2$ is nonnegative. Thus T^2 is quasihyponormal since

$$T^{2*}D_{T^2}T^2 = Q^2(R^4 - Q^4)Q^2 \oplus QR^2(R^2 - Q^2)R^2Q \oplus \bigoplus_{k=2}^{\infty} O \geq O$$

but not hyponormal once $R^4 - Q^4$ is not nonnegative.

Solution 9.16. Take an invertible operator T on a Hilbert space $\mathcal{H}$.

(a) If $O \leq T^*T - TT^*$, then $O \leq T^{-1}(T^*T - TT^*)T^{-1*} = T^{-1}T^*TT^{*-1} - I$. Thus $I \leq T^{-1}T^*TT^{*-1}$ and hence $T^*T^{-1}T^{*-1}T = (T^{-1}T^*TT^{*-1})^{-1} \leq I$. Therefore, $O \leq T^{-1*}(I - T^*T^{-1}T^{*-1}T)T^{-1} = T^{-1*}T^{-1} - T^{-1}T^{*-1}$. Outcome: T^{-1} is hyponormal whenever T is.

(b) Suppose T is paranormal. Then, for every x in $\mathcal{H}$,

$$\|x\|^2 = \|TT^{-1}x\|^2 \leq \|T^2(T^{-1}x)\| \|T^{-1}x\| = \|Tx\| \|T^{-1}x\|.$$

Take any y in $\mathcal{H} = \mathcal{R}(T)$ so that $y = Tx$, $x = T^{-1}y$ and $T^{-1}x = T^{-2}y$ for some x in $\mathcal{H}$. Thus $\|T^{-1}y\|^2 \leq \|y\| \|T^{-2}y\|$ by the above inequality, and so T^{-1} is paranormal.

Solution 9.17. *Claim.* Take any nonzero x in $\mathcal{H}$. If T is paranormal, then

$$\|Tx\|^n \leq \|T^n x\| \|x\|^{n-1} \quad \text{for every} \quad n \geq 1.$$

Proof. If T is paranormal, then $\|Tx\|^2 \leq \|T^2 x\| \|x\|$, and therefore

$$\|Tx\|^{2n} \le \|T(Tx)\|^n \|x\|^n$$

for every $n \ge 1$, for every $x \in \mathcal{H}$. Suppose the claimed inequality holds for some integer $n \ge 1$ so that

$$\|T(Tx)\|^n \le \|T^n(Tx)\| \|Tx\|^{n-1}$$

for every $x \in \mathcal{H}$ such that $Tx \ne 0$. Then

$$\|Tx\|^{n+1} \|Tx\|^{n-1} = \|Tx\|^{2n} \le \|T^n(Tx)\| \|Tx\|^{n-1} \|x\|^n,$$

and hence $\|Tx\|^{n+1} \le \|T^{n+1}x\| \|x\|^n$ for every $x \in \mathcal{H}$. Thus the claimed inequality holds for $n+1$ whenever it holds for n. Since it holds trivially for $n = 1$, the proof by induction is complete. $\qquad\square$

Solution 9.18. For any operator T in $\mathcal{B}[\mathcal{H}]$, $\lim_n \|T^n\|^{\frac{1}{n}} = r(T) \le \|T\|$ by the Gelfand–Beurling formula. If T is paranormal, then $\frac{\|Tx\|^n}{\|x\|^n} \le \frac{\|T^n x\|}{\|x\|}$ for every nonzero $x \in \mathcal{H}$ and every integer $n \ge 1$ (Problem 9.17). Therefore,

$$\|T\| = \sup_{x \ne 0} \frac{\|Tx\|}{\|x\|} \le \left(\sup_{x \ne 0} \frac{\|T^n x\|}{\|x\|} \right)^{\frac{1}{n}} = \|T^n\|^{\frac{1}{n}} \to r(T) \le \|T\|,$$

and hence $r(T) = \|T\|$; that is T is normaloid.

Solution 9.19. Let T^* be a hyponormal contraction on a Hilbert space $\mathcal{H}$. If $\mathcal{N}(A) = \mathcal{H}$, then $A = O$ is a trivial projection. Thus suppose $\mathcal{N}(A) \ne \mathcal{H}$ and take an arbitrary vector $y \in \mathcal{H}\backslash\mathcal{N}(A)$. Note that $\|T^n y\| \ne 0$ for all n (reason: if $T^{n_0} y = 0$ for some n_0, then $\|T^{*n} T^n y\| = 0$ for every $n \ge n_0$ so that $Ay = \lim_n T^{*n} T^n y = 0$). Since hyponormal operators are paranormal, Problems 7.3 and 9.17 ensure that

$$\left(\frac{\|T^{n+1} y\|}{\|T^n y\|} \right)^n \le \left(\frac{\|T^* T^n y\|}{\|T^n y\|} \right)^n \le \frac{\|T^{*n} T^n y\|}{\|T^n y\|} \le 1$$

for every integer $n \ge 1$ whenever T^* is a hyponormal contraction. It is clear that $\lim_n \|T^{*n} T^n y\| = \|Ay\|$ by the very definition of A, viz. $T^{*n} T^n \xrightarrow{s} A$. Problem 6.3(b) says that $\lim_n \|T^n y\| = \|A^{\frac{1}{2}} y\|$. Moreover, these limits are nonzero because $y \notin \mathcal{N}(A)$. Hence

$$\lim_n \frac{\|T^{*n} T^n y\|}{\|T^n y\|} = \frac{\|Ay\|}{\|A^{\frac{1}{2}} y\|}.$$

Now recall that, if a decreasing sequence of positive numbers $\{\beta_n\}$ converges to a nonzero limit, then $\limsup_n \left(\frac{\beta_{n+1}}{\beta_n} \right)^n = 1$. Thus, as T is a contraction,

$$\limsup_n \left(\frac{\|T^{n+1} y\|}{\|T^n y\|} \right)^n = 1.$$

Therefore,

$$1 \le \frac{\|Ay\|}{\|A^{\frac{1}{2}}y\|} \le 1;$$

that is, $\|A^{\frac{1}{2}}y\| = \|Ay\|$. But $O \le A \le I$ (Problem 6.1), and so $O \le A - A^2$ (Problem 2.3). Then take the square root $O \le (A - A^2)^{\frac{1}{2}}$ and compute

$$\|(A - A^2)^{\frac{1}{2}}y\|^2 = \langle(A - A^2)y \, ; y\rangle = \|A^{\frac{1}{2}}y\|^2 - \|Ay\|^2 = 0$$

so that $Ay = A^2y$. (Indeed, $(A - A^2)y = (A - A^2)^{\frac{1}{2}}(A - A^2)^{\frac{1}{2}}y = 0$.) Since y is an arbitrary vector in $\mathcal{H}\backslash\mathcal{N}(A)$, and since $Ax = A^2x$ holds trivially for every $x \in \mathcal{N}(A)$, it follows that $A = A^2$.

Solution 9.20. If T is a contraction for which A is a projection, then

$$T = G \oplus S_+ \oplus U,$$

where G is strongly stable, S_+ is a unilateral shift, U is unitary, and any of the above direct summands may be missing (Problem 6.9). If T is cohyponormal then the above decomposition holds (Problem 9.19) and, in this case, the direct summand S_+ must be missing because it is not cohyponormal (S_+ is nonnormal hyponormal, and direct summands of hyponormal operators are again hyponormal). If T is completely nonunitary, then the direct summand U also is missing, and therefore $T = G$: strongly stable.

Solution 9.21. Let T be a quasinormal contraction. Problem 7.2(c) says that $(T^{*n}T^n)T = T(T^{*n}T^n)$ for every $n \ge 0$, and hence $AT = TA$ (reason: $T^{*n}T^n \xrightarrow{s} A$) so that A is a projection by Problem 6.2(c). If T is a quasinormal contraction, then it is a hyponormal contraction, and so A_* is a projection by Problem 9.19.

10
Proper Contractions

Let $\mathcal{X}$ and $\mathcal{Y}$ be normed spaces. Recall that a bounded linear transformation T in $\mathcal{B}[\mathcal{X}, \mathcal{Y}]$ is a contraction if $\|T\| \leq 1$ (i.e., $\|Tx\| \leq \|x\|$ for every x in $\mathcal{X}$), and a strict contraction if $\|T\| < 1$ (i.e., $\sup_{x \neq 0}(\|Tx\|/\|x\|) < 1$). We say that T is a *proper contraction* if $\|Tx\| < \|x\|$ for every nonzero vector x in $\mathcal{X}$. Note that the concepts of proper and strict contractions make sense only if $\mathcal{X} \neq \{0\}$. It is clear that

$$\|T\| < 1 \quad \Longrightarrow \quad \|Tx\| < \|x\| \text{ for every } 0 \neq x \in \mathcal{X} \quad \Longrightarrow \quad \|T\| \leq 1.$$

As we shall see next, in a Hilbert space these implications are equivalent to

$$T^*T \prec I \quad \Longrightarrow \quad T^*T < I \quad \Longrightarrow \quad T^*T \leq I.$$

Problem 10.1. Take $T \in \mathcal{B}[\mathcal{X}, \mathcal{Y}]$, where $\mathcal{X}$ is a Hilbert space and $\mathcal{Y}$ is an inner product space. Let I be the identity operator on $\mathcal{X}$ and show that

(a) T is a contraction if and only if $T^*T \leq I$,

(b) T is a proper contraction if and only if $T^*T < I$,

(c) T is a strict contraction if and only if $T^*T \prec I$.

These classes are related by proper inclusion (e.g., the identity is a trivial example of nonproper contraction):

$$\text{Strict Contraction} \subset \text{Proper Contraction} \subset \text{Contraction}.$$

Problem 10.2. Exhibit a diagonal nonstrict proper contraction.

However, *the concepts of proper and strict contractions coincide for compact operators.*

Problem 10.3. Let $\mathcal{X}$ be a Hilbert space and $\mathcal{Y}$ an inner product space. Prove that a compact proper contraction in $\mathcal{B}[\mathcal{X}, \mathcal{Y}]$ is a strict contraction.

If T is a contraction or a strict contraction on a Hilbert space, then so is T^* (because $\|T^*\| = \|T\|$). This also happens for proper contractions.

Problem 10.4. Take T in $\mathcal{B}[\mathcal{H}, \mathcal{K}]$, where $\mathcal{H}$ and $\mathcal{K}$ are Hilbert spaces. Show that T is a proper contraction if and only if T^* is a proper contraction.

Problem 10.5. Use the inequality of Problem 9.17 to prove the following proposition. *A strongly stable paranormal operator is a proper contraction.*

If T in $\mathcal{B}[\mathcal{H}]$ is a strongly stable contraction, then it is usual to say that T is a C_0.-contraction (or a contraction of class C_0.). If T^* is a strongly stable contraction, then T is a $C_{\cdot 0}$-contraction (or a contraction of class $C_{\cdot 0}$). On the other extreme, if a contraction T on a Hilbert space $\mathcal{H}$ is such that $T^n x \not\to 0$ for every nonzero x in $\mathcal{H}$, then it is said to be a C_1.-contraction (or a contraction of class C_1.). Dually, if a contraction T is such that $T^{*n} x \not\to 0$ for every nonzero x in $\mathcal{H}$, then it is a $C_{\cdot 1}$-contraction (or a contraction of class $C_{\cdot 1}$). These are the Nagy–Foiaş classes of contractions [52, p. 72]. All combinations are possible leading to classes C_{00}, C_{01}, C_{10} and C_{11}. Note that T and T^* are both strongly stable if and only if T is of class C_{00}.

Problem 10.6. Exhibit contractions of classes C_{00}, C_{01}, C_{10} and C_{11}, and prove the following assertions.

  (a) Every uniformly stable contraction is of class C_{00}.

  (b) Every compact proper contraction is of class C_{00}.

  (c) There exists a compact contraction of class C_{00} that is not proper.

  (d) Every contraction not in class C_{00} is normaloid and nonstrict.

  (e) There exists a normaloid nonstrict proper contraction of class C_{00}.

Remarks: It is well known that *if a contraction has no nontrivial invariant subspace, then it is either a C_{00}, a C_{01}, or a C_{10}-contraction* (see e.g., [31, p. 71]). A classical open question asks whether this can be sharpened to C_{00} only: *does a contraction not in C_{00} have a nontrivial invariant subspace?* (see [30] for equivalent versions of it). Although the above question remains unanswered, a recent result [27] ensures that *if a contraction not in C_{00} has no nontrivial invariant subspace, then there is no nonzero quasinilpotent operator in its commutant.* This prompts the question: what is left of its commutant?

Problem 10.7. Every proper contraction is weakly stable. Prove.

In general, strongly stable contractions and proper contractions are not related. There exist strongly stable contractions that are not proper (even a nilpotent contraction may not be proper — see Solution 10.6(c)), and there exist proper contractions that are not strongly stable.

Problem 10.8. Show that the weighted unilateral shift on ℓ_+^2,

$$T_+ = \text{shift}\left(\left\{\frac{\sqrt{(k+1)(k+3)}}{k+2}\right\}_{k=0}^{\infty}\right),$$

is a hyponormal nonstrict proper contraction that is not strongly stable. Indeed, T_+ is a hyponormal nonstrict proper contraction of class C_{10}.

Problem 10.9. Show that the weighted unilateral shift of Problem 10.8 is, in fact, subnormal; a subnormal nonstrict proper contraction of class C_{10}.

Strongly stable paranormal operators are proper contractions (Problem 10.5). The converse fails even for subnormal operators (Problem 10.9) but holds for quasinormal operators. Apply the decomposition of Problem 6.9 together with Problems 9.21 and 10.5 to prove the next proposition.

Problem 10.10. If T is a quasinormal operator, then the following assertions are pairwise equivalent.

(a) T is a proper contraction.

(b) T is a C_{00}-contraction.

(c) T is strongly stable.

Thus *the concepts of strong stability and proper contractiveness coincide for quasinormal operators.* Here is another proof that does not use Problem 10.5, and hence does not pass through paranormal operators.

Problem 10.11. Take $T \in \mathcal{B}[\mathcal{H}]$ and prove the following assertions.

(a) T is a proper contraction if and only if T^*T is a proper contraction.

(b) If T is a contraction, then it is a proper contraction if and only if the point spectrum of T^*T is included in the interval $[0,1)$.

(c) If a normaloid operator T is strongly stable but not a proper contraction, then $|T|$ is not strongly stable.

Now apply assertion (c) to supply a direct proof (this time without using Problem 10.5 as we did in Solution 10.10) for the following statement.

(d) A strongly stable quasinormal operator is a proper contraction.

Take an arbitrary operator T on a Hilbert space $\mathcal{H}$ with $\dim \mathcal{H} > 1$. Set

$$\mathcal{M} = \{x \in \mathcal{H}: \|Tx\| = \|T\|\|x\|\}.$$

We know from Problem 7.5 that $\mathcal{M}$ is a subspace of $\mathcal{H}$ and, if T is hyponormal, then $\mathcal{M}$ is T-invariant. Show that this can be extended from hyponormal to paranormal operators and, consequently, conclude that *nonproper paranormal contractions have a nontrivial invariant subspace* [15].

Problem 10.12. Prove the following assertions.

(a) If T is paranormal, then $\mathcal{M}$ is an invariant subspace for T.

(b) If there exists a paranormal contraction without a nontrivial invariant subspace, then it is a proper contraction.

Remarks: Besides being a proper contraction, if a *hyponormal* contraction T has no nontrivial invariant subspace, then its self-commutator D_T is a strict contraction [36] and so is the nonnegative operator $|T^2| - |T|^2$ [14]. Moreover, the strong limit A of $\{T^{*n}T^n\}$ is either null or a nonstrict proper contraction [36] with no eigenvalues [25] (also see [26]).

Solutions

Solution 10.1. Let $T^* \in \mathcal{B}[\mathcal{Y}, \mathcal{X}]$ be the adjoint of $T \in \mathcal{B}[\mathcal{X}, \mathcal{Y}]$, where $\mathcal{X}$ is a Hilbert space and $\mathcal{Y}$ is an inner product space. Take an arbitrary vector x in $\mathcal{X}$. Recall that $\|Tx\|^2 = \langle Tx ; Tx \rangle = \langle T^*Tx ; x \rangle$ and $\|x\|^2 = \langle x ; x \rangle$.

(a) $\|Tx\| \le \|x\|$ if and only if $\langle T^*Tx ; x \rangle \le \langle x ; x \rangle$; equivalently, $T^*T \le I$.

(b) $\|Tx\| < \|x\|$ for $x \neq 0$ if and only if $\langle T^*Tx ; x \rangle < \langle x ; x \rangle$ for $x \neq 0$; which means $T^*T < I$.

(c) Since $\|Tx\|^2 \le \|T\|^2\|x\|^2$ we get $\langle T^*Tx ; x \rangle \le \|T\|^2\langle x ; x \rangle$, and hence $T^*T \le \|T\|^2 I$ or, equivalently, $T^*T - I \le (\|T\|^2 - 1)I$. If $\|T\| < 1$, then $1 - \|T\|^2 > 0$ so that $(1 - \|T\|^2)I \succ O$ because $I \succ O$, which implies $(\|T\|^2 - 1)I \prec O$. Therefore, $T^*T - I \prec O$; that is, $T^*T \prec I$. Conversely, if $T^*T \prec I$; that is, if $I - T^*T \succ O$, then there exists $\gamma > 0$ such that $\gamma\|x\|^2 \le \langle (I - T^*T)x ; x \rangle = \|x\|^2 - \|Tx\|^2$. Therefore, $\|Tx\|^2 \le (1 - \gamma)\|x\|^2$ so that $\|T\|^2 < 1 - \gamma < 1$, and hence $\|T\| < 1$.

Solution 10.2. Consider the diagonal operator $D = \mathrm{diag}\left(\left\{\frac{k+1}{k+2}\right\}_{k=0}^{\infty}\right)$ of Solution 3.1(d). It is a proper contraction in $\mathcal{B}[\ell_+^p]$ which is not a strict contraction. Indeed, $D = \mathrm{diag}\left(\frac{1}{2}, \frac{2}{3}, \frac{3}{4}, \ldots\right)$ is such that $\|Dx\| < \|x\|$ for every nonzero x in ℓ_+^p and $\|D\| = \sup_k \frac{k+1}{k+2} = 1$, and therefore D is a nonstrict proper contraction.

Solution 10.3. Take any T in $\mathcal{B}[\mathcal{X}, \mathcal{Y}]$, where $\mathcal{X}$ is a Hilbert space and $\mathcal{Y}$ is an inner product space. If T is a contraction, then $\|T^*\| = \|T\| \leq 1$ so that $\|T^*Tx\| \leq \|T^*\| \|Tx\| \leq \|Tx\|$ for every x in $\mathcal{X}$. Therefore, if T is a proper contraction, then so is T^*T. Thus $T^*Tx = \lambda x$ implies $|\lambda| \|x\| < \|x\|$ for every nonzero x in $\mathcal{X}$ so that $\sigma_P(T^*T)$ is included in the open unit disc. If T is compact, then T^*T is compact ($\mathcal{B}_\infty[\mathcal{X}]$ is a two-sided ideal of $\mathcal{B}[\mathcal{X}]$), and therefore $\sigma(T^*T)\backslash\{0\} = \sigma_P(T^*T)\backslash\{0\}$ (Fredholm alternative). Then $\sigma(T^*T)$ also is included in the open unit disc so that $r(T^*T) < 1$ (since the spectrum is closed); but $\|T\|^2 = r(T^*T)$. Hence T is a strict contraction.

Solution 10.4. Take any T in $\mathcal{B}[\mathcal{H}, \mathcal{K}]$, where $\mathcal{H}$ and $\mathcal{K}$ are Hilbert spaces. Recall that $\|T^*x\|^2 = \langle TT^*x \, ; x \rangle \leq \|TT^*x\| \|x\|$ for every $x \in \mathcal{H}$. Suppose T is a proper contraction and take any nonzero x in $\mathcal{H}$. If $T^*x \neq 0$, then $\|TT^*x\| < \|T^*x\| \neq 0$ so that $\|T^*x\|^2 < \|T^*x\| \|x\|$. Hence $\|T^*x\| < \|x\|$. If $T^*x = 0$, then $\|T^*x\| < \|x\|$ trivially. Thus T^* is a proper contraction whenever T is. The converse holds since $T^{**} = T$.

Solution 10.5. If T in $\mathcal{B}[\mathcal{H}]$ is paranormal, then $\frac{\|Tx\|^n}{\|x\|^n} \leq \frac{\|T^nx\|}{\|x\|}$ for every $x \neq 0$ in $\mathcal{H}$ and every $n \geq 1$ (Problem 9.17). Take any $x \neq 0$. If $\|T^nx\| \to 0$, then we get $\left(\frac{\|Tx\|}{\|x\|}\right)^n \to 0$ so that $\frac{\|Tx\|}{\|x\|} < 1$, and T is a proper contraction.

Solution 10.6. Trivial examples of $\mathcal{C}_{00}$, $\mathcal{C}_{01}$, $\mathcal{C}_{10}$ and $\mathcal{C}_{11}$-contractions are the null operator, a backward unilateral shift, a unilateral shift and the identity operator, respectively. Now consider the results of Problem 3.3.

(a) Uniform stability is $*$-invariant and implies strong stability.

(b) A compact proper contraction is a strict contraction (Problem 10.3). Thus, if T is a compact proper contraction, then $r(T) \leq \|T\| < 1$, which implies that T is uniformly stable (Problem 8.6). But uniformly stable contractions are of class $\mathcal{C}_{00}$ (assertion (a)).

(c) A nilpotent contraction on a finite-dimensional space is clearly compact (since it acts on a finite-dimensional space) and uniformly stable (since it is nilpotent), and hence of class $\mathcal{C}_{00}$ (assertion (a)). For instance, $\left(\begin{smallmatrix} 0 & 0 \\ 1 & 0 \end{smallmatrix}\right)$ in $\mathcal{B}[\mathbb{C}^2]$ is a nilpotent nonproper contraction on $\mathbb{C}^2$.

(d) A contraction T not in $\mathcal{C}_{00}$ is not uniformly stable by assertion (a), and hence $1 \leq r(T) \leq \|T\| \leq 1$ according to Problem 8.6.

(e) The diagonal operator $D = \text{diag}\left(\left\{\frac{k+1}{k+2}\right\}_{k=0}^\infty\right)$ in $\mathcal{B}[\ell_+^2]$ is a normaloid nonstrict proper contraction of class $\mathcal{C}_{00}$ (it is self-adjoint and strongly stable but not uniformly stable — see Solutions 3.1(d) and 10.2).

Solution 10.7. Every proper contraction is completely nonisometric (that is, if $\|Tx\| < \|x\|$ for every nonzero x in $\mathcal{H}$, then there is no nonzero reducing subspace $\mathcal{M}$ for T such that $\|Tu\| = \|u\|$ for all u in $\mathcal{M}$), and so completely nonunitary. Thus, by Problem 6.10, a proper contraction is weakly stable.

Solution 10.8. Consider a weighted unilateral shift $T_+ = \text{shift}(\{\alpha_k\}_{k=0}^\infty)$ on ℓ_+^2 for some bounded weight sequence $\{\alpha_k\}_{k=0}^\infty \in \ell_+^\infty$. Take an arbitrary $x = \{\xi_k\}_{k=0}^\infty = (\xi_0, \xi_1, \xi_2, \ldots)$ in ℓ_+^2. It is easy to show by induction that

$$T_+^n x = \left(0, \ldots, 0, \ \prod_{j=0}^{n-1}\alpha_j \xi_0, \ \prod_{j=1}^{n}\alpha_j \xi_1, \ \prod_{j=2}^{n+1}\alpha_j \xi_2, \ \ldots, \ \prod_{j=k}^{n+k-1}\alpha_j \xi_k, \ \ldots \right),$$

where the initial entries (from position 0 to position $n-1$) are all zero and the entry at the position $n+k$ is $\prod_{j=k}^{n+k-1}\alpha_j \xi_k$ for every $k \geq 0$,

$$T_+^{*n} x = \left(\prod_{j=0}^{n-1}\alpha_j \xi_n, \ \prod_{j=1}^{n}\alpha_j \xi_{n+1}, \ \prod_{j=2}^{n+1}\alpha_j \xi_{n+2}, \ \ldots, \ \prod_{j=k}^{n+k-1}\alpha_j \xi_{n+k}, \ \ldots \right),$$

and so

$$\|T_+^n x\|^2 = \sum_{k=0}^\infty \prod_{j=k}^{n+k-1} |\alpha_j|^2 |\xi_k|^2 \quad \text{and} \quad \|T_+^{*n} x\|^2 = \sum_{k=0}^\infty \prod_{j=k}^{n+k-1} |\alpha_j|^2 |\xi_{n+k}|^2,$$

for every $n \geq 1$. Now suppose

(i) $\sup_{k\geq 0} |\alpha_k| \leq 1$.

Since $\|T_+\| = \sup_{k\geq 0} |\alpha_k|$ (Problem 5.9), T_+ is a contraction. In particular,

(ii) $\sup_{k\geq 0} |\alpha_k| = 1$

implies that T_+ is a nonstrict contraction. Moreover, if

(iii) $|\alpha_k| < 1$ for every $k \geq 0$,

then T_+ is a proper contraction because, in this case,

$$\|T_+ x\|^2 = \sum_{k=0}^\infty |\alpha_k|^2 |\xi_k|^2 < \sum_{k=0}^\infty |\xi_k|^2 = \|x\|^2$$

for every nonzero x. Furthermore, (i) also implies that

$$\|T_+^{*n} x\|^2 \leq \left(\sup_{k\geq 0} |\alpha_k|^2\right)^n \sum_{k=0}^\infty |\xi_{n+k}|^2 \leq \sum_{k=n}^\infty |\xi_k|^2$$

for each $n \geq 1$ so that $\|T_+^{*n} x\| \to 0$ as $n \to \infty$ for every x. Observe that

$$\|T_+^n x\|^2 \geq \left(\inf_{n\geq 1} \inf_{k\geq 0} \prod_{j=k}^{n+k-1} |\alpha_j|^2\right) \|x\|^2$$

for all $n \geq 1$ and every x. Thus, if

(iv) $\inf_{n\geq 1} \inf_{k\geq 0} \prod_{j=k}^{n+k-1} |\alpha_j|^2 > 0$,

then $\|T_+^n x\| \not\to 0$ as $n \to \infty$ for every $x \neq 0$. Therefore, if (i) and (iv) hold, then T_+ is a $\mathcal{C}_{10}$-contraction. Finally, recall from Problem 9.7 that if

(v) $0 \leq \alpha_k \leq \alpha_{k+1}$ for every $k \geq 0$,

then T_+ is hyponormal. Now, for each $k \geq 0$, set

$$\alpha_k = \frac{\sqrt{(k+1)(k+3)}}{k+2} \quad \text{so that} \quad \prod_{j=k}^{n+k-1} |\alpha_j|^2 = \frac{(k+1)(n+k+2)}{(k+2)(n+k+1)}$$

for every $n \geq 1$, which is readily verified by induction. This sequence satisfies all the above properties (i) to (v). Indeed, for every $k \geq 0$ and $n \geq 1$,

$$0 < \alpha_k^2 = \frac{k^2+4k+3}{k^2+4k+4} = 1 - \frac{1}{(k+2)^2} < 1 \quad \text{and} \quad \prod_{j=k}^{n+k-1} |\alpha_j|^2 > \frac{k+1}{k+2} \geq \frac{1}{2}.$$

Solution 10.9. If $\alpha_k = \frac{\sqrt{(k+1)(k+3)}}{k+2}$ for every $k \geq 0$, then the weighted unilateral shift $T_+ = \text{shift}(\{\alpha_k\}_{k=0}^{\infty})$ on ℓ_+^2 is injective (with positive weights) and $\|T_+\| = 1$ (since the positive weight sequence $\{\alpha_k\}_{k=0}^{\infty}$ is increasing and $\lim_k \alpha_k = 1$). Let λ be the measure on $[0,1]$ given by $d\lambda(x) = x\,d\mu(x)$, where μ stands for the Lebesgue measure on $[0,1]$. Write dx for $d\mu(x)$, as usual. Consider the measure $\nu = \lambda + \frac{1}{2}\delta_1$ on $[0,1]$, where δ_1 is the unit point measure concentrated at 1. This ν is a probability measure on $[0,1]$ that contains 1 in its support (indeed, $[\nu] = [0,1]$ and $\int d\nu(x) = \int x\,dx + \frac{1}{2} = 1$). Recall that $\prod_{k=0}^{n-1} \alpha_k^2 = \frac{n+2}{2(n+1)} = \frac{1}{2(n+1)} + \frac{1}{2}$ for every $n \geq 1$ (Solution 10.8). Thus

$$\int x^{2n}\,d\nu(x) = \int x^{2n} x\,dx + \frac{1}{2} = \frac{1}{2(n+1)} + \frac{1}{2} = \prod_{k=0}^{n-1} \alpha_k^2,$$

for every $n \geq 1$, and hence T_+ is subnormal (compare with Solution 7.14).

Solution 10.10. If T is a quasinormal contraction, then A and A_* are projections (Problem 9.21). But we know from Problem 6.9 that if a contraction T is such that A and A_* are projections, then

$$T = B \oplus S_- \oplus S_+ \oplus U,$$

where B is a C_{00}-contraction, S_+ is a unilateral shift, S_- is a backward unilateral shift, U is unitary, and any of the above direct summands may be missing. If, in addition, T is a proper contraction, then any direct summand of it is again a proper contraction, and hence $T = B$, a C_{00}-contraction (reason: S_+, S_-^* and U are isometries, thus nonproper contractions, and proper contractiveness is $*$-invariant by Problem 10.4). Thus (a) implies (b). Tautologically, (b) implies (c) for every operator. Since quasinormal operators are paranormal, (c) implies (a) by Problem 10.5.

Solution 10.11. Let T be an operator acting on a Hilbert space $\mathcal{H}$.

(a) $\|Tx\|^2 = \langle T^*Tx\,; x\rangle \leq \|T^*Tx\|\|x\| \leq \|T\|\|Tx\|\|x\|$ for every x in $\mathcal{H}$. Thus T is a proper contraction if and only if T^*T is.

(b) If T is a proper contraction, then its point spectrum $\sigma_P(T)$ is included in the open unit disc, and so is the point spectrum $\sigma_P(T^*T)$ of T^*T by assertion (a). Since $\sigma(T^*T) \subseteq [0, \infty)$ (cf. Problem 8.3(e)), we may conclude: if T is a proper contraction, then $\sigma_P(T^*T) \subseteq [0, 1)$. Conversely, if T is a nonproper contraction, then there exists $x \neq 0$ in $\mathcal{H}$ such that $\|Tx\| = \|x\|$, and therefore $T^*Tx = x$ by Problem 7.5 so that $1 \in \sigma_P(T^*T)$, and hence $\sigma_P(T^*T) \not\subseteq [0, 1)$. Equivalently, if T is a contraction and $\sigma_P(T^*T) \subseteq [0, 1)$, then T is a proper contraction. Outcome: if T is a contraction, then it is a proper contraction if and only if the point spectrum of T^*T is included in the interval $[0, 1)$.

(c) Recall that

$$T^n \overset{s}{\longrightarrow} O \implies T^n \overset{w}{\longrightarrow} O \implies \sup_n \|T^n\| < \infty \implies r(T) \leq 1.$$

Thus, if T is strongly stable, then $r(T) \leq 1$. A normaloid is an operator T for which $r(T) = \|T\|$. If T is strongly stable and normaloid, then $r(T^*T) = \|T^*T\| = \|T\|^2 = r(T)^2 \leq 1$. If T is not a proper contraction, then assertion (b) ensures that $1 \in \sigma_P(T^*T)$ so that there exists $x \neq 0$ in $\mathcal{H}$ for which $\|(T^*T)^n x\| = \|x\| \neq 0$ for every integer $n \geq 1$, and hence $(T^*T)^n \overset{s}{\nrightarrow} O$. Therefore, if a normaloid operator T is strongly stable but not a proper contraction, then $T^*T = |T|^2$ is not strongly stable, and so $|T|$ is not strongly stable (Problem 3.2(b)).

(d) Now suppose T is quasinormal so that it is normaloid (Problems 7.4 and 7.10). If it is strongly stable but not a proper contraction, then assertion (c) says that $|T|$ is not strongly stable, but this implies that T itself is not strongly stable (cf. Problem 7.2(e)), which is a contradiction. Conclusion: If T is quasinormal and strongly stable, then it must be a proper contraction.

Solution 10.12. Consider the subspace $\mathcal{M} = \{x \in \mathcal{H}: \|Tx\| = \|T\|\|x\|\}$.

(a) If T is a paranormal operator on $\mathcal{H}$, that is, if $\|Tx\|^2 \leq \|T^2x\|\|x\|$ for every $x \in \mathcal{H}$, then $\mathcal{M}$ is T-invariant. Indeed, if $x \in \mathcal{M}$, then

$$\|T^2x\|\|x\| \leq \|T\|\|Tx\|\|x\| = \|Tx\|^2 \leq \|T^2x\|\|x\|,$$

and hence $\|T(Tx)\| = \|T\|\|Tx\|$ so that $Tx \in \mathcal{M}$. Thus $T(\mathcal{M}) \subseteq \mathcal{M}$.

(b) Take a contraction T on $\mathcal{H}$ so that $\|Tx\| \leq \|x\|$ for every x in $\mathcal{H}$. If T is a strict contraction, then it is a proper contraction. If T is a nonstrict contraction, then $\|T\| = 1$ and so $\mathcal{M} = \{x \in \mathcal{H}: \|Tx\| = \|x\|\}$. If T is a nonstrict paranormal contraction and has no nontrivial invariant subspace, then the invariant subspace $\mathcal{M}$ is trivial: either $\mathcal{M} = \{0\}$ or $\mathcal{M} = \mathcal{H}$; but if $\mathcal{M} = \mathcal{H}$, then T is an isometry (Problem 2.5), and isometries have nontrivial invariant subspaces whenever $\dim \mathcal{H} > 1$ (Problem 8.14). Thus $\mathcal{M} = \{0\}$ so that $\|Tx\| < \|x\|$ for every nonzero x in $\mathcal{H}$; that is, T is a proper contraction.

11

Quasireducible Operators

An operator T on a complex Hilbert space is reducible if and only if there exists a nonscalar operator L such that $LT = TL$ and $T^*L = LT^*$ (Problem 8.16). That is, if and only if there exists a nonscalar L in $\{T\}' \cap \{T^*\}'$, where $\{T\}'$ denotes the commutant of T. Equivalently, T is reducible if and only if both T and T^* lie in $\{L\}'$ for some nonscalar operator L. Quasireducibility was recently introduced in [34].

Definition: An operator T on a complex Hilbert space $\mathcal{H}$ is *quasireducible* if there exists a nonscalar operator L on $\mathcal{H}$ such that

$$LT = TL \quad \text{and} \quad \text{rank}\left((T^*L - LT^*)T - T(T^*L - LT^*)\right) \leq 1.$$

Note that nonscalar operators exist only on spaces of dimension greater than 1, and so the concept of quasireducible (and reducible) are germane to operators on Hilbert spaces of dimension greater than 1. Recall that *rank* means dimension of range, and the only operator with rank 0 is the null operator. Also recall that by a *commutator* we mean an operator C such that $C = AB - BA$ for some operators A and B, usually denoted by $[A, B]$. Thus an operator T is quasireducible if there exists a nonscalar L in $\{T\}'$ such that either $T^*L - LT^*$ lies in $\{T\}'$ (equivalently, if the commutator $[(T^*L - LT^*), T] = (T^*L - LT^*)T - T(T^*L - LT^*)$ has rank 0) or the commutator $[(T^*L - LT^*), T]$ has rank 1. It is clear that *every reducible operator is quasireducible*. Indeed, an operator T is reducible if and only if there exists a nonscalar L such that $LT = TL$ and $T^*L - LT^* = O$, which trivially implies $(T^*L - LT^*)T - (T^*L - LT^*) = O$.

Here is a necessary and sufficient condition for quasireducibility that will be used throughout this chapter. Let $D_T = [T^*, T] = T^*T - TT^*$ be the self-commutator of an operator T on a complex Hilbert space.

Problem 11.1. Show that T is quasireducible if and only if there exists a nonscalar L such that

$$LT = TL \quad \text{and} \quad \text{rank}\,(D_T L - L D_T) \leq 1.$$

Problem 11.2. Prove: If T is quasireducible, then

(a) λT is quasireducible for every $\lambda \in \mathbb{C}$,

(b) $\lambda I + T$ is quasireducible for every $\lambda \in \mathbb{C}$,

(c) T^* is quasireducible.

Note that $D_{N+T} = D_N + D_T + 2\,\text{Re}\,(T^*N - NT^*)$. If N is a normal operator in $\{T\}'$, then the Fuglede–Putnam Theorem ensures that N^* lies in $\{T\}'$, and hence $N \in \{T^*\}'$. Therefore, $D_{N+T} = D_T$ and $D_T N = N D_T$. Thus it might be tempting to generalize item (b) in Problem 11.2 as follows. *If T is quasireducible, then $N + T$ is quasireducible for every normal N that commutes with T.* This, however, does not add anything to Problem 11.2(b). Indeed, if N is a nonscalar normal in $\{T\}'$, then T and $N + T$ are reducible (Problem 8.16), and hence trivially quasireducible; if N is scalar, the above italicized result collapses to Problem 11.2(b).

Recall that an operator T is nilpotent if $T^n = O$ for some positive integer n. The least integer n such that $T^n = O$ is the *nilpotence index* of T.

Problem 11.3. Let T be a nilpotent operator acting on a Hilbert space of dimension greater than 1. Prove the following assertion. If T is nilpotent of index $n+1$ for some $n \geq 1$, then either T^n is reducible, or T is quasireducible with nilpotence index 2 on a two-dimensional space.

Problem 11.4. Show that every nilpotent operator of index 2 is quasireducible. In fact, a nilpotent operator of index 2 acting on a Hilbert space of dimension greater than 2 is reducible; on a two-dimensional space it is irreducible but quasireducible.

Problem 11.5. Exhibit a nilpotent operator of index greater than 2 that is not quasireducible.

Problem 11.6. Show that product and (ordinary) sum of quasireducible operators are not necessarily quasireducible.

Question: Is the square of a quasireducible operator quasireducible?

Problem 11.7. Every operator unitarily equivalent to a quasireducible operator is quasireducible. That is, show that quasireducibility is preserved under unitary equivalence.

Problem 11.8. Both reducibility and quasireducibility are not preserved under similarity. Exhibit a nonquasireducible operator that is similar to a reducible operator.

Problem 11.9. A unilateral shift of multiplicity 1 is quasireducible.

Problem 11.10. If a quasinormal operator is not a multiple of an isometry, then it is reducible. Prove.

Use the von Neumann–Wold decomposition to prove the next assertion.

Problem 11.11. The only irreducible quasinormal operator is a unilateral shift of multiplicity 1.

Corollary: Every quasinormal operator is quasireducible.

We say that a diagonal operator has *multiplicity* 1 if the diagonal sequence is made up of distinct elements.

Problem 11.12. Show that every injective unilateral weighted shift whose self-commutator has multiplicity 1 is not quasireducible.

Problem 11.13. Exhibit a subnormal operator that is not quasireducible.

The equivalent condition of Problem 11.1 played a central role so far. It points to an apparent gap between the classes of reducible and quasireducible operators. Consider the class $\mathcal{C}$ of all operators T for which there exists a nonscalar L such that

$$LT = TL \quad \text{and} \quad D_T L = L D_T.$$

It is clear that $\mathcal{C}$ includes the class of all reducible operators and is included in the class of all quasireducible operators:

$$\text{Reducible} \subseteq \mathcal{C} \subset \text{Quasireducible}.$$

In fact, the second inclusion is proper: there exist quasireducible operators not in $\mathcal{C}$. That is, operators T such that $\text{rank}\,(D_T L - L D_T) = 1$ for some nonscalar L in $\{T\}'$ but $\text{rank}\,(D_T L - L D_T) \neq 0$ for all nonscalar L in $\{T\}'$ (samples: a unilateral shift of multiplicity 1 or, simply, $\left(\begin{smallmatrix} 0 & 1 \\ 0 & 0 \end{smallmatrix}\right)$ on $\mathbb{C}^2$). What about the first inclusion? Is it an equality? Is it true that if T is irreducible, then $\text{rank}\,(D_T L - L D_T) \geq 1$ for every nonscalar L in $\{T\}'$?

Question: Does $\mathcal{C}$ coincide with the class of all reducible operators?

There are many ways to reformulate this question as we shall see below. First let us recall that a *commutator* on a normed space $\mathcal{X}$ is an operator C on $\mathcal{X}$ such that $C = ST - TS$, where both S and T are operators on $\mathcal{X}$.

Problem 11.14. Take S and T in $\mathcal{B}[\mathcal{X}]$, $\mathcal{X}$ is a normed space. Show that

(a) $ST^{n+1} - T^{n+1}S = \sum_{k=0}^{n} T^k (ST - TS) T^{n-k}$ for each integer $n \geq 0$.

Use item (a) to prove that nonzero commutators are nonscalar. That is,

(b) the only scalar commutator is the null operator.

Apply the result of Problem 11.14(b) to frame the next proposition.

Problem 11.15. Let $\mathcal{H}$ be a Hilbert space. Show that $T \in \mathcal{B}[\mathcal{H}]$ is reducible if and only if there exits a nonscalar $L \in \mathcal{B}[\mathcal{H}]$ such that

(a) $LT = TL, \quad D_T L = L D_T \quad$ and

(b) $(T^* D_L - D_L T^*)T = T(T^* D_L - D_L T^*)$.

Thus the above question can be rewritten as follows. Can we drop assertion (b) from the statement of Problem 11.15? Equivalently (see Solution 11.15), does there exist a nonscalar normal in $\{T\}'$ whenever there exists a nonscalar L such that both T and D_T lie in $\{L\}'$? Here is another way to look upon the same question. Let $\mathcal{A}_T$ be the unital algebra of all operators that commute with T and with D_T,

$$\mathcal{A}_T = \{L \in \mathcal{B}[\mathcal{H}]: LT = TL \text{ and } D_T L = L D_T\} = \{T\}' \cap \{D_T\}',$$

and let $\mathbb{C}$ denote the trivial unital algebra of all scalar operators so that

$$\mathbb{C} \subseteq \mathcal{A}_T \subseteq \{T\}'.$$

Consider the following well-known result (see e.g., [45, p. 4] or [22, p. 130]):

If $C = ST - TS$ and $CT = TC$, then C is *quasinilpotent*.

Problem 11.16. Show that the assertions below are pairwise equivalent.

(a) T is normal.

(b) $D_T T = T D_T$.

(c) $\mathcal{A}_T = \{T\}'$.

$\mathcal{A}_T = \{T\}'$ if and only if T is normal (Problem 11.16). On the other hand, if $\mathcal{A}_T = \mathbb{C}$, then T is irreducible. Indeed, if T is reducible then there exists a nonscalar in $\{T\}' \cap \{T^*\}'$ (Problem 8.16), which clearly lies in $\mathcal{A}_T$. The converse, however, holds if and only if the previous question has an affirmative answer. Is it true that $\mathcal{A}_T = \mathbb{C}$ for every irreducible T?

Solutions

Solution 11.1. Let T and L be operators acting on the same Hilbert space. If $LT = TL$, then

$$
\begin{aligned}
D_T L - L D_T &= (T^*T - TT^*)L - L(T^*T - TT^*) \\
&= T^*LT - TT^*L - LT^*T + TLT^* \\
&= (T^*L - LT^*)T - T(T^*L - LT^*).
\end{aligned}
$$

Thus T is quasireducible if and only if there exists a nonscalar L such that

$$
LT = TL \quad \text{and} \quad \operatorname{rank}(D_T L - L D_T) \leq 1.
$$

Solution 11.2. For any operator S let $\{S\}'$ denote the commutant of S and let D_S be the self-commutator of S. Take any λ in $\mathbb{C}$ and note that

$$
\{\lambda T\}' = \{\lambda I + T\}' = \{T\}'.
$$

Moreover, since

$$
D_{\lambda T} = \bar{\lambda} T^* \lambda T - \lambda T \bar{\lambda} T^* = \bar{\lambda}\lambda(T^*T - TT^*) = |\lambda|^2 D_T
$$

and

$$
\begin{aligned}
D_{\lambda I + T} &= (\bar{\lambda} I + T^*)(\lambda I + T) - (\lambda I + T)(\bar{\lambda} I + T^*) \\
&= |\lambda|^2 I + \bar{\lambda} T + \lambda T^* + T^*T - |\lambda|^2 I - \lambda T^* - \bar{\lambda} T - TT^* \\
&\quad - T^*T - TT^* = D_T,
\end{aligned}
$$

it follows that

$$
D_{\lambda T} L - L D_{\lambda T} = |\lambda|^2 (D_T L - L D_T) \quad \text{and} \quad D_{\lambda I + T} L - L D_{\lambda I + T} = D_T L - L D_T
$$

for every operator L, and hence (a) and (b) hold (i.e., λT and $\lambda I + T$ are quasireducible whenever T is quasireducible). Indeed, if L is a nonscalar operator such that $LT = TL$ and $\operatorname{rank}(D_T L - L D_T) \leq 1$, then

$$
L(\lambda T) = (\lambda T)L, \quad L(\lambda T + I) = (\lambda T + I)L, \quad \text{and}
$$

$$
\operatorname{rank}(D_{\lambda T} L - L D_{\lambda T}) \leq \operatorname{rank}(D_{\lambda I + T} L - L D_{\lambda I + T}) \leq 1.
$$

To prove (c) observe that (i) $D_{T^*} = -D_T$, (ii) D_S is self-adjoint for every S, (iii) L^* is a nonscalar operator in $\{T^*\}'$ whenever L is a nonscalar operator in $\{T\}'$, and recall that (iv) $\operatorname{rank}(S) = \operatorname{rank}(S^*)$ for every S (see e.g., [32, p. 427]). Therefore, if there exists a nonscalar L such that $LT = TL$ and $\operatorname{rank}(D_T L - L D_T) \leq 1$, then L^* is a nonscalar such that $L^*T^* = T^*L^*$ and

$$
D_{T^*} L^* - L^* D_{T^*} = L^* D_T - D_T L^* = (D_T L)^* - (L D_T)^* = (D_T L - L D_T)^*,
$$

and hence T^* is quasireducible whenever T is quasireducible because

$$\text{rank}(D_{T^*}L^* - L^*D_{T^*}) = \text{rank}((D_TL - LD_T)^*) = \text{rank}(D_TL - LD_T).$$

Solution 11.3. Take an arbitrary operator $T \neq O$ on a Hilbert space $\mathcal{H}$ so that $\mathcal{N}(T) \neq \mathcal{H}$. Since $\mathcal{N}(T)$ is an invariant subspace for T, and since $T|_{\mathcal{N}(T)}$ is trivially null, it follows that T can be written, with respect to the decomposition $\mathcal{H} = \mathcal{N}(T) \oplus \mathcal{N}(T)^\perp$, as (see Problem 4.2)

$$T = \begin{pmatrix} T|_{\mathcal{N}(T)} & X \\ O & Y \end{pmatrix} = \begin{pmatrix} O & X \\ O & Y \end{pmatrix} \quad \text{and hence} \quad T^n = \begin{pmatrix} O & XY^{n-1} \\ O & Y^n \end{pmatrix}$$

for every integer $n \geq 1$ (by induction), where

$$X : \mathcal{N}(T)^\perp \to \mathcal{N}(T) \quad \text{and} \quad Y : \mathcal{N}(T)^\perp \to \mathcal{N}(T)^\perp$$

are bounded linear transformations (recall: $Y^0 = I$ for every operator Y, where I is the identity operator). Suppose T is nilpotent of index $n+1$. That is, suppose $T^{n+1} = O$ and $T^n \neq O$ for some integer $n \geq 1$. Then

$$Y^n = O \quad \text{and} \quad XY^{n-1} \neq O.$$

Indeed, if $Y^n \neq O$, then there exists $v \neq 0$ in $\mathcal{N}(T)^\perp$ such that $v = Y^n u$ for some $u \neq 0$ in $\mathcal{N}(T)^\perp$, and hence $Xv = XY^n u$ and $Yv = Y^{n+1}u$. Since $T^{n+1} = O$, it follows that $XY^n = O$ and $Y^{n+1} = O$. Thus $Xv = Yv = 0$, and so $T(0 \oplus v) = Xv \oplus Yv = 0$. That is, $0 \oplus v$ lies in $\mathcal{N}(T)$. But this is a contradiction (the only vector $0 \oplus v$ in $\mathcal{N}(T) \oplus \{0\}$ is $0 = 0 \oplus 0$). Hence $Y^n = O$, which implies that $XY^{n-1} \neq O$ because $T^n \neq O$. Therefore,

$$T^n = \begin{pmatrix} O & Z \\ O & O \end{pmatrix} \quad \text{with} \quad Z = XY^{n-1} : \mathcal{N}(T)^\perp \to \mathcal{N}(T).$$

Remarks: Clearly, $\dim \mathcal{H} \geq 2$ (otherwise there is no nilpotent operator of positive index acting on $\mathcal{H}$). Moreover, $\mathcal{N}(T) \neq \{0\}$ because 0 is an eigenvalue of T (if $T^{n+1} = TT^n = O$ and $T^n \neq O$, then there exists a nonzero vector x in $\mathcal{H}$ such that $T^n x \neq 0$ and $T(T^n x) = 0$). Since $\mathcal{N}(T)$ is a proper subspace of $\mathcal{H}$, it follows that $\mathcal{N}(T)$ is nontrivial so that $\mathcal{N}(T)$ and $\mathcal{N}(T)^\perp$ are both nonzero. (In fact, $\mathcal{N}(T)$ is a nontrivial invariant subspace for T, which can likewise be verified by Problem 1.3; also see Problem 1.5).

With respect to the same decomposition $\mathcal{H} = \mathcal{N}(T) \oplus \mathcal{N}(T)^\perp$, set

$$Q = \begin{pmatrix} ZZ^* & O \\ O & Z^*Z \end{pmatrix} \quad \text{so that} \quad QT^n = T^nQ = \begin{pmatrix} O & ZZ^*Z \\ O & O \end{pmatrix}.$$

If the nonnegative Q is nonscalar, then T^n is reducible (by Problem 8.16, since nonnegative operators are, obviously, normal). If Q is scalar, then $Z^*Z = \lambda I_\perp$ and $ZZ^* = \lambda I$ for some positive λ, where I and $I_\perp$ stand for the identity operator on $\mathcal{N}(T)$ and $\mathcal{N}(T)^{\perp}$, respectively. (Recall: $Z \neq O$ since $T^n \neq O$, and both Z^*Z and ZZ^* are nonnegative.) Thus $\lambda^{-\frac{1}{2}} Z$ is an isometry and a coisometry; that is, $\lambda^{-\frac{1}{2}} Z$ is a unitary operator. Therefore, $\mathcal{N}(T)$ and $\mathcal{N}(T)^{\perp}$ are unitarily equivalent, and hence $\dim \mathcal{N}(T) = \dim \mathcal{N}(T)^{\perp}$. Take any operator $A \colon \mathcal{N}(T) \to \mathcal{N}(T)$ and set, still on $\mathcal{H} = \mathcal{N}(T) \oplus \mathcal{N}(T)^{\perp}$,

$$N = \begin{pmatrix} A & O \\ O & \lambda^{-1} Z^* A Z \end{pmatrix} \quad \text{so that} \quad N T^n = T^n N = \begin{pmatrix} O & AZ \\ O & O \end{pmatrix}.$$

If $\dim \mathcal{N}(T) \geq 2$, then let A be any nonscalar normal operator so that N is a nonscalar normal operator as well and, again, T^n is reducible. On the other hand, if $\dim \mathcal{N}(T) = 1$, then $\dim \mathcal{H} = 2$. In this case we may assume that $T^n = \begin{pmatrix} 0 & \xi \\ 0 & 0 \end{pmatrix}$ on $\mathbb{C}^2$ for some $\xi \neq 0$ in $\mathbb{C}$, which implies $n = 1$. That is,

T is a nilpotent operator of index 2 on a two-dimensional space.

Indeed, if $T^n = \begin{pmatrix} 0 & \xi \\ 0 & 0 \end{pmatrix}$, then $T = \begin{pmatrix} \alpha & \beta \\ 0 & \alpha \end{pmatrix}$, and hence $T^n = \begin{pmatrix} \alpha^n & \xi \\ 0 & \alpha^n \end{pmatrix}$ by induction, for some $\alpha, \beta \in \mathbb{C}$ (reason: T commutes with T^n). Thus $\alpha = 0$ so that $T^2 = O$, and therefore $n = 1$. Finally, note that any nonscalar operator L that commutes with $T = \begin{pmatrix} 0 & \xi \\ 0 & 0 \end{pmatrix}$ is of the form $L = \begin{pmatrix} \alpha & \beta \\ 0 & \alpha \end{pmatrix}$ with $\beta \neq 0$, which is never normal. Then it follows by Problem 8.16 that T is irreducible; but quasireducible because $D_T L - L D_T = \begin{pmatrix} 0 & -2\beta \\ 0 & 0 \end{pmatrix}$.

Solution 11.4. Let $\mathcal{H}$ be a Hilbert space such that $\dim \mathcal{H} \geq 2$, and let T be an idempotent operator of index $n+1$ for some integer $n > 1$ acting on $\mathcal{H}$. From Solution 11.3 we know that T^n is not reducible only if $\dim \mathcal{H} = 2$ and, in such a case, $n = 1$ and T is quasireducible. Therefore, if T is nilpotent of index 2 (i.e., if $n = 1$), then either T is reducible if $\dim \mathcal{H} > 2$, or irreducible but quasireducible if $\dim \mathcal{H} = 2$.

Solution 11.5. Identify operators on $\mathbb{C}^3$ with their matrices with respect to the canonical basis for $\mathbb{C}^3$. Take

$$T = \begin{pmatrix} 0 & 1 & 1 \\ 0 & 0 & 1 \\ 0 & 0 & 0 \end{pmatrix} \quad \text{so that} \quad D_T = \begin{pmatrix} -2 & -1 & 0 \\ -1 & 0 & 1 \\ 0 & 1 & 2 \end{pmatrix}$$

in $\mathcal{B}[\mathbb{C}^3]$. T is a nilpotent of index 3 ($T^2 \neq O$ and $T^3 = O$) that is not quasireducible. In fact, any L that commutes with T is of the form

$$L = \begin{pmatrix} \alpha & \beta & \gamma \\ 0 & \alpha & \beta \\ 0 & 0 & \alpha \end{pmatrix} \quad \text{so that} \quad D_T L - L D_T = \begin{pmatrix} \beta & -2\beta - \gamma & -2\beta - 4\gamma \\ 0 & -2\beta & -2\beta - \gamma \\ 0 & 0 & \beta \end{pmatrix},$$

for arbitrary α, β and γ in $\mathbb{C}$. Hence $\operatorname{rank}(D_T L - L D_T) \geq 2$ whenever L is nonscalar (i.e., whenever β and γ are not both null).

Solution 11.6. As before, 3×3 matrices will be enough. Take

$$T = \begin{pmatrix} 0 & 1 & 0 \\ 0 & 0 & 1 \\ 0 & 0 & 0 \end{pmatrix} \quad \text{so that} \quad T^2 = \begin{pmatrix} 0 & 0 & 1 \\ 0 & 0 & 0 \\ 0 & 0 & 0 \end{pmatrix}.$$

T^2 is nilpotent of index 2 on $\mathbb{C}^3$, and hence reducible according to Problem 11.4. T is quasireducible (since $\operatorname{rank}(T^2 D_T - D_T T^2) = 1$). Thus $I + T$ is quasireducible by Problem 11.2(b). However, $T(I + T) = T + T^2$, which is both a product and a sum of quasireducible operators, is not quasireducible as we saw in Solution 11.5.

Solution 11.7. Let $\mathcal{H}$ and $\mathcal{K}$ be unitarily equivalent Hilbert spaces. Take a quasireducible operator T in $\mathcal{B}[\mathcal{H}]$. Let L be a nonscalar operator in $\mathcal{B}[\mathcal{H}]$ such that $LT = TL$ and $\operatorname{rank}(D_T L - L D_T) \leq 1$. Suppose S in $\mathcal{B}[\mathcal{K}]$ is unitarily equivalent to T. Let $U : \mathcal{K} \to \mathcal{H}$ be a unitary transformation such that $S = U^* T U$. Consider the operator $M = U^* L U$ in $\mathcal{B}[\mathcal{K}]$. It is readily verified that M is nonscalar (because L is nonscalar) and commutes with S (since L commutes with T). Moreover,

$$D_S M - M D_S = U^*(D_T L - L D_T)U$$

so that $D_S M - M D_S$ and $D_T L - L D_T$ are unitarily equivalent. (It is worth noticing that $D_S M - M D_S$ and $D_T L - L D_T$ might not be similar if S was just similar to T.) But rank is invariant under unitary equivalence. Indeed,

$$\begin{aligned} U(\mathcal{R}(D_S M - M D_S)) &= U(D_S M - M D_S)(\mathcal{K}) \\ &= (D_T L - L D_T) U(\mathcal{K}) \\ &= (D_T L - L D_T)(\mathcal{H}) = \mathcal{R}(D_T L - L D_T) \end{aligned}$$

so that $\mathcal{R}(D_S M - M D_S)$ and $\mathcal{R}(D_T L - L D_T)$ are isomorphic linear spaces, and hence have the same (finite) dimension. Since $\operatorname{rank}(D_T L - L D_T) \leq 1$, it follows that $\operatorname{rank}(D_S M - M D_S) \leq 1$. Thus S is quasireducible.

Solution 11.8. Again, identify operators on $\mathbb{C}^3$ with 3×3 matrices. Take

$$T = \begin{pmatrix} 1 & -1 & 1 \\ 0 & 0 & 0 \\ 0 & 1 & 0 \end{pmatrix} \quad \text{so that} \quad D_T = \begin{pmatrix} -2 & -1 & 2 \\ -1 & 2 & -1 \\ 2 & -1 & 0 \end{pmatrix}.$$

Any nonscalar L that commutes with T is of the form

$$L = \begin{pmatrix} \alpha & \beta & \gamma \\ 0 & \alpha - \gamma & 0 \\ 0 & -\beta & \alpha - \gamma \end{pmatrix}$$

where β and γ cannot be both null. Thus

$$D_T L - L D_T = \begin{pmatrix} \beta - 2\gamma & 2\gamma - 6\beta & \beta - 4\gamma \\ -\gamma & 0 & -\gamma \\ 2\gamma - \beta & 4\beta & 2\gamma - \beta \end{pmatrix},$$

and hence $\operatorname{rank}(D_T L - L D_T) \geq 2$ for every nonscalar L that commutes with T. Outcome: T is not quasireducible. Now put

$$\tilde{T} = \begin{pmatrix} 1 & 0 & 0 \\ 0 & 0 & 0 \\ 0 & 1 & 0 \end{pmatrix} \quad \text{and} \quad W = \begin{pmatrix} 1 & 0 & 1 \\ 0 & 1 & 0 \\ 0 & 0 & 1 \end{pmatrix},$$

so that W is invertible and $WT = \tilde{T}W$. Thus the reducible $\tilde{T} = 1 \oplus \left(\begin{smallmatrix} 0 & 0 \\ 1 & 0 \end{smallmatrix} \right)$ is similar to T, which is not even quasireducible.

Solution 11.9. Let S_+ be the canonical unilateral shift of multiplicity 1 on $\ell_+^2 = \ell_+^2(\mathbb{C})$. Recall that $I - S_+ S_+^* = P_0 = \operatorname{diag}(e_0) = \operatorname{diag}(1, 0, 0, \ldots)$ (see Solution 5.6), and also that S_+ is an isometry (i.e., $S_+^* S_+ = I$). Hence

$$(S_+^* S_+ - S_+ S_+^*) S_+ - S_+ (S_+^* S_+ - S_+ S_+^*) = -S_+ (I - S_+ S_+^*) = -S_+ P_0,$$

which is a rank-1 operator ($S_+ P_0 x = \xi_0 e_1$ for each $x = \{\xi_k\}_{k=0}^\infty$ in ℓ_+^2), and therefore S_+ is quasireducible. Thus, according to Problems 5.3 and 11.7, *every* unilateral shift of multiplicity 1 is quasireducible.

Solution 11.10. A (linear) isometry acting on a (complex nonzero) Hilbert space $\mathcal{H}$ is precisely an operator V such that $V^* V = I$. Let T be an operator on $\mathcal{H}$. T is a multiple of an isometry if and only if the nonnegative $T^* T$ is scalar. If a quasinormal operator T is not a multiple of an isometry, then the nonnegative (thus normal) $T^* T$ is nonscalar and commutes with T (by the definition of quasinormal), and hence T is reducible (Problem 8.16).

Solution 11.11. Let $\mathcal{H}$ be a complex Hilbert space, of dimension greater than 1 (otherwise there is no reducible or quasireducible operator on $\mathcal{H}$), and let T be a nonzero quasinormal operator on $\mathcal{H}$ (the zero operator on $\mathcal{H}$ is trivially reducible). If T is not a multiple of an isometry, then it is reducible according to Problem 11.10. If T is a multiple of an isometry, then the von Neumann–Wold decomposition says that

$$T = \alpha S_+ \oplus \alpha U$$

for some nonzero scalar α, where S_+ is a unilateral shift of arbitrary multiplicity and U is unitary. If both direct summands are present, then T is trivially reducible; if $T = \alpha U$, then it is normal so that it is again reducible (Problem 8.16). Thus suppose $T = \alpha S_+$. If S_+ has multiplicity greater than 1, then it is reducible (Problem 5.5) so that T is, once again, reducible. If S_+ has multiplicity 1, then it is irreducible but quasireducible (Problems 5.6 and 11.9), and so is T according to Problem 11.2.

Solution 11.12. According to Problems 5.10 and 11.7 there is no loss of generality in assuming unilateral weighted shifts with nonnegative weights. Thus let $T_+ = \text{shift}(\{\omega_k\}_{k=1}^\infty)$ be a unilateral weighted shift on ℓ_+^2 with nonnegative weights $\{\omega_k\}_{k=1}^\infty$ so that $T_+^* T_+ = \text{diag}(\omega_1^2, \omega_2^2, \omega_3^2, \omega_4^2, ...)$ and $T_+ T_+^* = \text{diag}(0, \omega_1^2, \omega_2^2, \omega_3^2, ...)$, and hence the self-commutator of T_+ is

$$D_{T_+} = T_+^* T_+ - T_+ T_+^* = \text{diag}(\{\delta_k\}_{k=1}^\infty),$$

whose diagonal entries are $\delta_1 = \omega_1^2$ and $\delta_{k+1} = \omega_{k+1}^2 - \omega_k^2$ for every $k \geq 1$. Recall: T_+ is injective if and only if $\omega_k \neq 0$ for every $k \geq 1$ (Problem 5.9). Let A be an arbitrary operator on ℓ_+^2. Identify A with the infinite matrix $[\alpha_{j,k}]_{j,k \geq 1}$ that represents it with respect to the canonical basis for ℓ_+^2.

Claim 1. If a diagonal operator $D = \text{diag}(\{\lambda_k\}_{k=1}^\infty)$ has multiplicity 1, then A commutes with D if and only if A is a diagonal operator.

Proof. It is readily verified that

$$DA - AD = \left[\alpha_{j,k}(\lambda_j - \lambda_k)\right]_{j,k \geq 1}.$$

If $\lambda_j \neq \lambda_k$ for $j \neq k$, then $DA = AD$ if and only if $\alpha_{j,k} = 0$ for $j \neq k$. $\square$

Claim 2. If A commutes with an injective unilateral weighted shift T_+, then A is lower triangular (i.e., all entries above the main diagonal are zero) with a constant main diagonal. Moreover, the entries of the lower diagonals (i.e., of each diagonal below the main diagonal) are either all zero or all nonzero.

Proof. For each $j \geq 1$ consider the sequences

$$UR(A)_j = \{\alpha_{j,k}\}_{k=j}^\infty,$$

referred to as an *upper row* of A (i.e., a row of $[\alpha_{j,k}]_{j,k \geq 1}$ starting at the main diagonal of $[\alpha_{j,k}]_{j,k \geq 1}$), and

$$SUR(A)_j = \{\alpha_{j,k}\}_{k=j+1}^\infty,$$

referred to as a *strictly upper row* of A (i.e., a row of $[\alpha_{j,k}]_{j,k \geq 1}$ starting just after the main diagonal of $[\alpha_{j,k}]_{j,k \geq 1}$). Take an arbitrary integer $j \geq 1$. If $T_+ A = A T_+$, then

$$UR(T_+ A)_j = 0 \quad \text{implies} \quad UR(A T_+)_j = 0.$$

If $\omega_k \neq 0$ for every $k \geq 1$, then

$$UR(A T_+)_j = 0 \quad \text{implies} \quad SUR(A)_j = 0,$$

and

$$SUR(A)_j = 0 \quad \text{implies} \quad UR(T_+ A)_{j+1} = 0.$$

Therefore, since $UR(T_+ A)_1 = 0$, a trivial induction shows that

$$SUR(A)_j = 0 \quad \text{for every} \quad j \geq 1.$$

That is, A is lower triangular. Moreover, $T_+A = AT_+$ also implies that

$$\omega_{j+i}\,\alpha_{j+i,j} = \omega_j\,\alpha_{j+i+1,j+1}$$

for every $j \geq 1$ and $i \geq 0$. If $\omega_k \neq 0$ for every $k \geq 1$, then

$$\alpha_{j+i+1,j+1} = \frac{\omega_{j+i}}{\omega_j}\alpha_{j+i,j}$$

for every $j \geq 1$ and $i \geq 0$. For $i = 0$ we get $\alpha_{j+1,j+1} = \alpha_{j,j}$, which means that A has a constant main diagonal. For any $i \geq 1$ this ensures that either the sequence $\{\alpha_{j+i,j}\}_{j=1}^{\infty}$ is null or $\alpha_{j+i,j} \neq 0$ for all $j \geq 1$. That is, the ith lower diagonal of A is either zero or entirely made of nonzero entries.   □

Claim 3. If a lower triangular A with lower diagonals either zero or entirely nonzero does not commute with a diagonal operator D of multiplicity 1, then $DA - AD$ is not finite-rank.

Proof. Take the canonical basis $\{e_k\}_{k=1}^{\infty}$ for ℓ_+^2 (i.e., $e_k = (0, \dots, 0, 1, 0, \dots)$ for each $k \geq 1$, with the only nonzero entry equal to 1 at the kth position). Now recall that $DA - AD = [\alpha_{j,k}(\lambda_j - \lambda_k)]_{j,k \geq 1}$ for any diagonal operator $D = \text{diag}(\{\lambda_k\}_{k=1}^{\infty})$. If A is lower triangular (i.e., $\alpha_{j,k} = 0$ whenever $j < k$), then $(DA - AD)e_k$ coincides with the kth column of the lower triangular $DA - AD$. Since A does not commute with D, it follows by Claim 1 that A is not diagonal: there exists $\alpha_{j,k} \neq 0$ for some pair (j,k) with $j > k \geq 2$. Consequently, the lower diagonal to which it belongs is entirely nonzero. Since D is a diagonal of multiplicity 1 (i.e., $\lambda_j \neq \lambda_k$ whenever $j \neq k$), it follows that the lower triangular $DA - AD$ has at least one lower diagonal made up of nonzero entries. Hence the closure of span $\{(DA - AD)e_k\}_{k=1}^{\infty}$, $\bigvee\{(DA - AD)e_k\}_{k=1}^{\infty}$, is an infinite-dimensional subspace of $\mathcal{R}(DA - AD)$ so that $DA - AD$ is not finite-rank.   □

Outcome: Let T_+ be an injective unilateral weighted shift whose self-commutator $D_{T_+} = T_+^*T_+ - T_+T_+^*$ has multiplicity 1. If A commutes with T_+ and with D_{T_+}, then A is scalar (by Claims 1 and 2). If A commutes with T_+ but does not commute with D_{T_+}, then $D_{T_+}A - AD_{T_+}$ is not finite-rank (by Claims 2 and 3).

Solution 11.13. The unilateral weighted shift $T_+ = \text{shift}(\{\alpha_k\}_{k=0}^{\infty})$ with $\alpha_k = \left(\frac{k+1}{k+2}\right)^{\frac{1}{2}}$ for each $k \geq 0$ is subnormal according to Problem 7.14. It is clearly injective ($\alpha_k \neq 0$ for every $k \geq 0$). Recall that $D_{T_+} = \text{diag}(\{\delta_k\}_{k \geq 0})$ with $\delta_0 = \alpha_0^2$ and $\delta_{k+1} = \alpha_{k+1}^2 - \alpha_k^2$ for every $k \geq 0$ (Problem 9.7). In this particular case we have $\delta_k = \frac{1}{(k+1)(k+2)}$ for each $k \geq 0$ so that the self-commutator D_{T_+} has multiplicity 1 ($\delta_j \neq \delta_k$ if $j \neq k$ since $\delta_{k+1} < \delta_k$ for every $k \geq 0$). Thus T_+ is not quasireducible by Problem 11.12.

Solution 11.14. Let S and T be linear transformations of a linear space $\mathcal{X}$ into itself. Put $C = ST - TS$.

(a) Take an arbitrary integer $n \geq 0$. Since $CT^n = ST^{n+1} - TST^n$ we get

$$ST^{n+1} = T(ST^n) + CT^n.$$

Thus a trivial induction shows that, for every $n \geq 1$,

$$ST^n = T^n(ST^0) + \sum_{k=0}^{n-1} T^k(CT^{n-1-k}).$$

That is, $ST^n - T^nS = \sum_{k=0}^{n-1} T^k CT^{n-1-k}$ for each integer $n \geq 1$.

(b) Let $\mathcal{X}$ be a normed space and suppose S and T are operators in $\mathcal{B}[\mathcal{X}]$. If $C = \gamma I$ for some scalar γ (i.e., if C is a scalar commutator), then

$$ST^n - T^nS = n\gamma T^{n-1} \quad \text{so that} \quad n|\gamma| \|T^{n-1}\| \leq 2\|S\| \|T\| \|T^{n-1}\|$$

for every $n \geq 1$. If $T^{n-1} \neq O$ and $T^n = O$ for some $n \geq 1$ (i.e., if either T is null or T is nilpotent of some index $n \geq 2$), then $n\gamma T^{n-1} = O$ so that $\gamma = 0$. If $T^n \neq O$ for every $n \geq 0$, then $n|\gamma| \leq 2\|S\| \|T\|$ for all $n \geq 1$, and hence $\gamma = 0$. Summing up: If C is scalar, then $C = O$.

Solution 11.15. If T is reducible, then (Problem 8.16) there exists a non-scalar L in $\{T\}' \cap \{T^*\}'$. Thus assertions (a) and (b) hold trivially. Conversely, take a nonscalar L and consider the commutator $C = T^*L - LT^*$. Recall from Solution 11.1 that assertion (a) is equivalent to $LT = TL$ and $CT = TC$. Hence, if (a) holds, then $D_C = C^*C - CC^*$ is given by

$$D_C = (L^*T - TL^*)C - C(L^*T - TL^*) = (L^*C - CL^*)T - T(L^*C - CL^*).$$

However, since $L^*T^* = T^*L^*$,

$$L^*C - CL^* = L^*(T^*L - LT^*) - (T^*L - LT^*)L^* = T^*D_L - D_LT^*.$$

Therefore, if assertion (b) also holds, then $D_C = O$; that is, C is normal. If C is nonscalar, then T is reducible (because $CT = TC$). If the commutator C is scalar, then $C = O$ by Problem 11.14(b) and hence the nonscalar L lies in $\{T\}' \cap \{T^*\}'$; that is, T is reducible (see Problem 8.16 again).

Solution 11.16. If T is normal (that is, if $D_T = O$), then $D_T T = TD_T$ and $\mathcal{A}_T = \{T\}'$ trivially. If $\{T\}' \subseteq \mathcal{A}_T$, then $T \in \mathcal{A}_T$ so that T commutes with D_T. Then $\mathcal{A}_T = \{T\}'$ implies $D_T T = TD_T$. Recall that $D_T = T^*T - TT^*$ is always self-adjoint, and hence normaloid. If $D_T T = TD_T$, then D_T is quasinilpotent. Thus $D_T = O$ (the only normaloid quasinilpotent is the null operator). Therefore $D_T T = TD_T$ implies that T is normal.

12

The Lomonosov Theorem

Compact operators (on a complex Banach space of dimension greater than 1) have a nontrivial invariant subspace; a nontrivial hyperinvariant subspace, actually, if it is nonscalar. The definitive result in this line is due to Lomonosov [40]: *An operator has a nontrivial invariant subspace if it commutes with a nonscalar operator that commutes with a nonzero compact operator.* In fact, *every nonscalar operator that commutes with a nonscalar compact operator* (itself, in particular) *has a nontrivial hyperinvariant subspace.* Recall that on an infinite-dimensional normed space the only scalar compact operator is the null operator; on a finite-dimensional normed space every operator is compact.

The Lomonosov Theorem is a remarkable breakthrough on the invariant subspace problem. The full version of it, mentioned above, can be split into two parts:

(i) *If an operator commutes with a nonzero compact operator, then it has a nontrivial invariant subspace;*

(ii) *if it is nonscalar, then it has a nontrivial hyperinvariant subspace.*

Part (i) is a slightly weaker version of the full statement, whose proof is known as Hilden's Proof of Lomonosov's Theorem [42]. The purpose of the next problem is to prove part (i).

Problem 12.1. Take a nonzero compact operator K on a complex Banach space $\mathcal{X}$ of dimension greater than 1. Let T be an operator in $\mathcal{B}[\mathcal{X}]$ that commutes with K. Suppose T has no nontrivial invariant subspace and prove the following assertions.

(a) K is quasinilpotent, and hence αK is uniformly stable for all $\alpha \in \mathbb{C}$.

Hint: Fredholm alternative for compact operators and the Beurling–Gelfand formula for the spectral radius.

(b) For each nonzero vector x in $\mathcal{X}$ and each nonempty open subset U of $\mathcal{X}$, there exists a nonzero polynomial p such that $p(T)x \in U$.

Hint: If an operator has no nontrivial invariant subspace, then every nonzero vector is cyclic.

Assertions (a) and (b) are proved by using eigenspaces and cyclic spaces, respectively, which are two important sources of invariant subspaces. Both (a) and (b) rely on the hypothesis that T has no nontrivial invariant subspace. The program is to show that this assumption, through assertions (a) and (b), leads to a contradiction. Take any vector x_0 in $\mathcal{X}$ such that the origin is not in the closed ball $B_1[x_0] = \{x \in \mathcal{X} : \|x - x_0\| \leq 1\}$ nor in the closure of its image under K;

$$0 \notin B_1[x_0] \quad \text{and} \quad 0 \notin K(B_1[x_0])^-.$$

Since $K \neq O$, this happens for any vector $x_0 \in \mathcal{X}$ such that $\|K\| < \|Kx_0\|$. Indeed, if 0 lies in $B_1[x_0]$, then $\|x_0\| \leq 1$, and so $\|Kx_0\| \leq \|K\|$. If 0 lies in the closure $K(B_1[x_0])^-$ of $K(B_1[x_0])$ (i.e., if 0 is a point of adherence of $K(B_1[x_0])$), then there exists a sequence $\{x_n\}$ in $B_1[x_0]$ such that $Kx_n \to 0$, and hence $\|Kx_0\| = \lim_n \|K(x_n - x_0)\| \leq \|K\| \limsup_n \|x_n - x_0\| \leq \|K\|$. Thus take any x_0 in $\mathcal{X}$ that satisfies the above properties and, for each nonzero polynomial p, set

$$U_p(x_0) = \{x \in \mathcal{X} : \|p(T)x - x_0\| < 1\}.$$

Show that assertion (b) implies assertion (c) below.

(c) $U_p(x_0)$ is open in $\mathcal{X}$ and every nonzero vector of $\mathcal{X}$ lies in $U_p(x_0)$ for some p.

Since $0 \notin K(B_1[x_0])^-$, show that assertion (c) implies the next one.

(d) There exists a finite set $\mathcal{P}$ of polynomials such that, if $x \in K(B_1[x_0])$, then $p(T)x$ lies in $B_1(x_0)$ for some $p \in \mathcal{P}$.

Finally, use assertion (d) to prove assertion (e).

(e) There is a sequence $\{p_k\}$ of polynomials in $\mathcal{P}$ such that, for each $n \geq 1$,

$$p_1(T) \dots p_n(T) K^n x_0 \in B_1[x_0].$$

Summing up: Assertion (b) ensures the existence of a sequence $\{x_n\}$ of vectors in $B_1[x_0]$ such that $x_n = p_1(T) \dots p_n(T) K^n x_0$ for every $n \geq 1$, with each p_k in $\mathcal{P}$, where $\mathcal{P}$ is a finite set of polynomials so that $\sup_{p \in \mathcal{P}} \|p(T)\|$

is finite. Put $\alpha = \sup_{p \in \mathcal{P}} \|p(T)\|$. On the other hand, assertion (a) ensures that $\|(\alpha K)^n\| \to 0$. Therefore,

$$\|x_n\| = \|p_1(T) \dots p_n(T) K^n x_0\| \leq \alpha^n \|K^n x_0\| \leq \|(\alpha K)^n\| \|x_0\| \to 0,$$

and the $B_1[x_0]$-valued sequence $\{x_n\}$ converges to 0. Since $B_1[x_0]$ is closed in $\mathcal{X}$, the Closed Set Theorem (cf. Solution 1.1) ensures that $0 \in B_1[x_0]$. But this contradicts the fact that $0 \notin B_1[x_0]$. Conclusion: *If an operator T commutes with a nonzero compact K, then T has a nontrivial invariant subspace*, which completes the proof of part (i) of the Lomonosov Theorem.

Besides very basic functional analysis, Hilden's Proof of Lomonosov's Theorem uses only two elementary results from operator theory, namely, the Fredholm alternative and the Beurling–Gelfand formula. A proof of the full version of the Lomonosov Theorem requires more than that.

First recall that the *convex hull* of a subset G of a linear space $\mathcal{X}$, denoted by $\mathrm{co}(G)$, is the intersection of all convex sets containing G; that is, $\mathrm{co}(G)$ is the smallest (in the inclusion ordering) convex set that contains G. Also recall that $\mathrm{co}(G)$ coincides with the set of all *convex linear combinations* of vectors in G; that is, $x \in \mathrm{co}(G)$ if and only if $x = \sum_{i=1}^{n} \alpha_i x_i$ for some finite set $\{x_i\}_{i=1}^{n}$ of vectors in G and some finite set of positive scalars $\{\alpha_i\}_{i=1}^{n}$ such that $\sum_{i=1}^{n} \alpha_i = 1$. Clearly, $\mathrm{co}(G)^- \subseteq \mathrm{co}(G^-)^-$ for every subset G of a normed space $\mathcal{X}$. A classical result on the geometry of Banach spaces is the Mazur Theorem.

Mazur Theorem. *The closure of the convex hull of every compact subset of a Banach space is compact.*

That is, if C is a compact set in a Banach space $\mathcal{X}$, then so is $\mathrm{co}(C)^-$. In fact, this can be readily verified by showing that $\mathrm{co}(C)$ is totally bounded whenever C is (see e.g., [8, p. 180]).

Now we borrow the notion of *compact mapping* from nonlinear functional analysis. Let D be a nonempty subset of a normed space $\mathcal{X}$. A mapping $F : D \to \mathcal{X}$ is *compact* if it is continuous and $F(B)^-$ is compact in $\mathcal{X}$ whenever B is a bounded subset of D. Recall that *a continuous image of any compact set is a compact set* (see e.g., [32, p. 149]). Thus, if D is a compact subset of $\mathcal{X}$, then every continuous mapping $F : D \to \mathcal{X}$ is compact. We shall, however, be concerned with the case where D (the domain of F) is not compact but is bounded. In this case, if F is continuous and $F(D)^-$ is a compact set, then F is a compact mapping (because $F(B)^- \subseteq F(D)^-$ whenever $B \subseteq D$). The central result required for proving the full version of the Lomonosov Theorem is the Schauder Fixed Point Theorem (see e.g., [8, p. 150]), which reads as follows.

Schauder Fixed Point Theorem. *Let D be a closed, bounded and convex subset of a normed space $\mathcal{X}$, and let $F: D \to \mathcal{X}$ be a compact mapping. If D is F-invariant, then F has a fixed point (i.e., if $F(D) \subseteq D$, then there exists $x \in D$ such that $F(x) = x$).*

Next take the algebra $\mathcal{B}[\mathcal{X}]$ of all operators on a normed space $\mathcal{X}$ and let $\mathcal{A}$ be a unital subalgebra of $\mathcal{B}[\mathcal{X}]$, which means that $\mathcal{A}$ is a linear manifold of $\mathcal{B}[\mathcal{X}]$ that contains the identity and is such that AB lies in $\mathcal{A}$ whenever A and B lie in $\mathcal{A}$. A subspace $\mathcal{M}$ of $\mathcal{X}$ is invariant for $\mathcal{A}$ (or is $\mathcal{A}$-invariant) if it is invariant for every operator in $\mathcal{A}$ (i.e., if $A(\mathcal{M}) \subseteq \mathcal{M}$ for every $A \in \mathcal{A}$). We say that $\mathcal{A}$ has no nontrivial invariant subspace if there is no nontrivial subspace of $\mathcal{X}$ that is invariant for every operator in $\mathcal{A}$. An intermediate stage towards a proof of the full version of the Lomonosov Theorem is the so-called Lomonosov Lemma.

Lomonosov Lemma. *Let K be a nonzero compact operator on a complex Banach space $\mathcal{X}$, and let $\mathcal{A}$ be a unital subalgebra of $\mathcal{B}[\mathcal{X}]$. If there is no nontrivial invariant subspace of $\mathcal{X}$ that is invariant for every operator in $\mathcal{A}$, then there exists L in $\mathcal{A}$ such that 1 is an eigenvalue of LK (i.e., such that $\mathcal{N}(I - LK) \neq \{0\}$).*

The purpose of the next problem is to prove the Lomonosov Lemma step by step. The first steps actually prepare the ground for an application of the Schauder Fixed Point Theorem.

Problem 12.2. Let $\mathcal{X}$ be a complex Banach space and let $\mathcal{A}$ be a unital subalgebra of $\mathcal{B}[\mathcal{X}]$ that has no nontrivial invariant subspace. To begin with, prove the following assertion.

(a) For each nonzero vector x in $\mathcal{X}$ and each nonempty open subset U of $\mathcal{X}$ there exists an operator A in $\mathcal{A}$ such that $Ax \in U$.

Let K be a nonzero compact operator on $\mathcal{X}$. Take $x_0 \in \mathcal{X}$ as in Problem 12.1 so that $0 \notin B_1[x_0]$ and $0 \notin K(B_1[x_0])^-$. For each operator A in $\mathcal{A}$ set

$$U_A(x_0) = \{x \in \mathcal{X}: \|Ax - x_0\| < 1\}.$$

Show that assertion (a) implies assertion (b) below.

(b) $U_A(x_0)$ is open in $\mathcal{X}$ and every nonzero vector of $\mathcal{X}$ lies in $U_A(x_0)$ for some operator A in $\mathcal{A}$.

Since $0 \notin K(B_1[x_0])^-$, show that assertion (b) implies the next one.

(c) There exists a finite subset $\mathcal{F}$ of $\mathcal{A}$ such that, if $x \in K(B_1[x_0])$, then Ax lies in $B_1(x_0)$ for some $A \in \mathcal{F}$.

For each A in $\mathcal{F}$ consider the function $\alpha_A: K(B_1[x_0]) \to \mathbb{R}$ given by

$$\alpha_A(x) = \max\{0, 1 - \|Ax - x_0\|\} \quad \text{for every} \quad x \in K(B_1[x_0]).$$

Take an arbitrary $x \in K(B_1[x_0])$. Assertion (c) ensures that there exists A in $\mathcal{F}$ such that $\|Ax - x_0\| < 1$, and hence $0 < \alpha_A(x)$ for some A in $\mathcal{F}$. Therefore, since $0 \leq \alpha_A(x)$ for every $x \in K(B_1[x_0])$ and each $A \in \mathcal{F}$, we get $0 < \sum_{A \in \mathcal{F}} \alpha_A(x) < \infty$ for every $x \in K(B_1[x_0])$. Put

$$\beta_A(x) = \frac{\alpha_A(x)}{\sum_{A \in \mathcal{F}} \alpha_A(x)} \quad \text{for each} \quad x \in K(B_1[x_0]),$$

which defines a function $\beta_A \colon K(B_1[x_0]) \to \mathbb{R}$. Now let $F \colon B_1[x_0] \to \mathcal{X}$ be a mapping defined by

$$F(x) = \sum_{A \in \mathcal{F}} \beta_A(Kx) A K x \quad \text{for every} \quad x \in B_1[x_0].$$

(d) Use the Mazur Theorem to prove that F is a compact mapping.

(e) Show that F is $B_1[x_0]$-invariant.

(f) Recall that $0 \notin B_1[x_0]$ and apply the Schauder Fixed Point Theorem to conclude that there exists L in $\mathcal{A}$ such that $\mathcal{N}(I - LK) \neq \{0\}$, thus completing the proof of the Lomonosov Lemma.

The proof of part (i) of the Lomonosov Theorem uses only elementary results of operator theory (cf. Solution 12.1). In order to prove part (ii), thus completing the full version of it, we shall apply the Lomonosov Lemma. The full version of the Lomonosov Theorem can be rephrased as follows.

Lomonosov Theorem. *If a nonscalar operator commutes with a nonzero compact operator, then it has a nontrivial hyperinvariant subspace.*

Take an operator T on a complex Banach space $\mathcal{X}$. Let $\{T\}'$ be the commutant of T, which is the unital subalgebra of $\mathcal{B}[\mathcal{X}]$ consisting of all operators in $\mathcal{B}[\mathcal{X}]$ that commute with T. Recall that a nontrivial hyperinvariant subspace for T is a nontrivial subspace of $\mathcal{X}$ that is invariant for every operator in $\{T\}'$.

Problem 12.3. Let T be an operator acting on a complex Banach space $\mathcal{X}$. Suppose there exists a nonzero compact operator K in $\{T\}'$, and suppose T has no nontrivial hyperinvariant subspace. Apply the Lomonosov Lemma to prove the following assertion.

(a) There exists an operator L in $\{T\}'$ such that $\mathcal{N}(I - LK)$ is nonzero and T-invariant.

Now use the above result to show that

(b) T has an eigenvalue (i.e., there is a $\lambda \in \mathbb{C}$ such that $\mathcal{N}(\lambda I - T) \neq \{0\}$).

But $\mathcal{N}(\lambda I - T)$ is a hyperinvariant subspace for T (Problem 8.2(a)). Therefore, if T has no nontrivial hyperinvariant subspace, then $\mathcal{N}(\lambda I - T) = \mathcal{X}$. Equivalently, $T = \lambda I$; that is, T is scalar. Summing up: *If an operator T has no nontrivial hyperinvariant subspace and commutes with a nonzero compact K, then T must be scalar*, proving the Lomonosov Theorem.

Recall that rank means dimension of range. Thus T commutes with K if and only if rank$(KT - TK) = 0$. Here is an extension of the Lomonosov Theorem. For simplicity, operators are assumed to act on a Hilbert space.

Problem 12.4. Let T, S and K be operators acting on a complex Hilbert space. Prove the following proposition.

(a) If $0 < \dim \mathcal{N}(S) < \infty$, $0 < \dim \mathcal{N}(S^*) < \infty$ and $\dim \mathcal{R}(TS-ST) = 1$, then $\sigma_P(T) \cup \sigma_P(T^*) \neq \varnothing$.

Now use it to prove the next one.

(b) If K is compact and $\dim \mathcal{R}(KT - TK) = 1$, then T has a nontrivial hyperinvariant subspace.

If $\dim \mathcal{R}(KT-TK) = 1$ (i.e., if rank$(KT-TK) = 1$), then K is nonzero and T is nonscalar. Therefore, combining the Lomonosov Theorem with the result of Problem 12.4(b) yields the following useful extension of it.

Extension of the Lomonosov Theorem. *If a nonscalar operator T is such that* rank$(KT-TK) \leq 1$ *for some nonzero compact operator K, then T has a nontrivial hyperinvariant subspace.*

Problem 12.4(b) was proved in a Hilbert space setting. Let us just mention that it holds in a Banach space setting as well (and so does the above extension of the Lomonosov Theorem). The proof of Problem 12.4(a), and hence the proof of Problem 12.4(b), in a complex Banach space $\mathcal{X}$ is based on the quotient space $\mathcal{X}/\mathcal{R}(S)^-$ rather than on the orthogonal complement $\mathcal{H} \ominus \mathcal{R}(S)^- = \mathcal{R}(S)^\perp = \mathcal{N}(S^*)$ (see [11], [28] and [29]).

Remark: The following open question is the so-called *hyperinvariant subspace problem*: Does every nonscalar operator have a nontrivial hyperinvariant subspace? It is worth noticing that if one could improve the above extension of the Lomonosov Theorem by replacing rank$(KT - TK) \leq 1$ with rank$(KT - TK) \leq 2$, then one would have solved affirmatively the hyperinvariant subspace problem. Indeed, for every operator T there exists a nonzero compact K such that rank$(KT - TK) \leq 2$ (reason: this always holds whenever K is a rank-1 operator).

If an operator has a compact self-commutator, then it is called *essentially normal*. Use the extension of the Lomonosov Theorem to prove the next assertion (in a complex Hilbert space of dimension greater than 1).

Problem 12.5. Every essentially normal quasireducible operator has a nontrivial invariant subspace.

Let $\{e_\gamma\}_{\gamma \in \Gamma}$ be an orthonormal basis for a Hilbert space $\mathcal{H}$ and let T be an operator on $\mathcal{H}$. If the family of nonnegative numbers $\{\langle |T|e_\gamma ; e_\gamma \rangle\}_{\gamma \in \Gamma}$ is summable, then its sum $\sum_{\gamma \in \Gamma} \langle |T|e_\gamma ; e_\gamma \rangle$ is independent of the choice of the orthonormal basis $\{e_\gamma\}_{\gamma \in \Gamma}$. An operator T in $\mathcal{B}[\mathcal{H}]$ is *trace-class* if the family $\{\langle |T|e_\gamma ; e_\gamma \rangle\}_{\gamma \in \Gamma}$ is summable (i.e., if $\sum_{\gamma \in \Gamma} \langle |T|e_\gamma ; e_\gamma \rangle < \infty$) for some orthonormal basis $\{e_\gamma\}_{\gamma \in \Gamma}$ (and hence for every orthonormal basis) for $\mathcal{H}$. Moreover, if T is trace-class, then $\{\langle Te_\gamma ; e_\gamma \rangle\}_{\gamma \in \Gamma}$ also is a summable family of scalars and the sum $\sum_{\gamma \in \Gamma} \langle Te_\gamma ; e_\gamma \rangle$ does not depend on the choice of the orthonormal basis $\{e_\gamma\}_{\gamma \in \Gamma}$. This sum is the *trace* of T, denoted by $\mathrm{tr}(T)$: if T is trace-class, then $\mathrm{tr}(T) = \sum_{\gamma \in \Gamma} \langle Te_\gamma ; e_\gamma \rangle$ for any orthonormal basis $\{e_\gamma\}_{\gamma \in \Gamma}$ for $\mathcal{H}$. Recall that every trace-class operator is compact (see e.g., [32, pp. 435–440]).

An operator $T \in \mathcal{B}[\mathcal{H}]$ is *finitely multicyclic* if there exists a *finite* set of vectors in $\mathcal{H}$, say $G = \{x_i\}_{i=1}^m$, such that

$$\bigvee \{f(T)x \in \mathcal{H} : x \in G \text{ and } f \in \mathrm{Rat}(\sigma(T))\} = \mathcal{H},$$

where $\mathrm{Rat}(\sigma(T))$ denotes the algebra of all rational functions with poles off the spectrum of T. In other words, T is finitely multicyclic if there exists a finite set G such that the linear span of $\{f(T)x : x \in G \text{ and } f \in \mathrm{Rat}(\sigma(T))\}$ is dense in $\mathcal{H}$. The vectors in G are called *generating vectors*. If $m \geq 1$ is the smallest number of generating vectors for a finitely multicyclic operator T, then T is called *m-multicyclic*. An important result on hyponormal operators is the Berger–Shaw Theorem [4], [5] (also see [9, p. 152]):

Berger–Shaw Theorem. *If a hyponormal operator T is m-multicyclic, then its self-commutator D_T is a trace-class operator and*

$$\mathrm{tr}(D_T) \leq \tfrac{m}{\pi} \mu(\sigma(T))),$$

where μ denotes the planar Lebesgue measure (i.e., $\mu(\sigma(T)) = \mathrm{area}(\sigma(T))$).

Show that this and Problem 12.5 are enough to guarantee that if there exists a hyponormal operator without a nontrivial invariant subspace, then it is not quasireducible.

Problem 12.6. Quasireducible hyponormal operators have a nontrivial invariant subspace.

Quasinormal operators are quasireducible (Problem 11.11) but there exist nonquasireducible subnormal operators (see Solution 11.13), and hence nonquasireducible hyponormal operators. The invariant subspace problem remains unsolved for hyponormal operators (see remarks that follow Problem 8.14). However, according to Problem 12.6, the invariant subspace problem for hyponormal operators is restricted to the class of nonquasireducible hyponormal operators.

The concept of quasireducibility is naturally connected with three major invariant subspace theorems, namely, Lomonosov's for compact operators, Berger–Shaw's for hyponormal operators and S. Brown's for subnormal operators (which says that *every subnormal operator has a nontrivial invariant subspace* — again, see remarks that follow Problem 8.14). Thus the fact that *a nonquasireducible subnormal operator has a nontrivial invariant subspace* is a trivial corollary of the S. Brown Theorem (recall: there exist nonquasireducible subnormal operators). However, an independent proof of the above italicized result would lead to a new proof for the S. Brown Theorem (via Problem 12.6), which would be a consequence of the Lomonosov and Berger–Shaw Theorems.

Solutions

Solution 12.1. Let K be a nonzero compact operator acting on a complex Banach space $\mathcal{X}$ of dimension greater than 1, which commutes with an operator T in $\mathcal{B}[\mathcal{X}]$. Suppose T has no nontrivial invariant subspace.

(a) If $T \in \mathcal{B}[\mathcal{X}]$ has no nontrivial invariant subspace and $\mathcal{X}$ has dimension greater than 1, then $\sigma_P(T) = \varnothing$ according to Problem 8.2(c). Take an arbitrary nonzero λ in $\mathbb{C}$. Since T commutes with $\lambda I - K$, it follows by Problem 8.2(a) that the subspace $\mathcal{N}(\lambda I - K)$ is T-invariant. If T has no nontrivial invariant subspace, then either $\mathcal{N}(\lambda I - K) = \mathcal{X}$ or $\mathcal{N}(\lambda I - K) = \{0\}$. If $\mathcal{N}(\lambda I - K) = \mathcal{X}$, then the nonzero compact K is scalar so that $\mathcal{X}$ is finite-dimensional (the identity is not compact on an infinite-dimensional space), and therefore we get $\sigma(T) = \sigma_P(T)$ (Problem 8.7), which contradicts the fact that $\sigma_P(T) = \varnothing$ (recall: $\sigma(T) \neq \varnothing$). Thus $\mathcal{N}(\lambda I - K) = \{0\}$ for every nonzero λ, and hence $\sigma_P(K) \subseteq \{0\}$. But K is a compact operator, and the Fredholm alternative says that $\sigma(K)\backslash\{0\} = \sigma_P(K)\backslash\{0\}$. Then $\sigma(K) = \{0\}$; that is, K is quasinilpotent ($r(K) = 0$) so that $r(\alpha K) = \alpha r(K) = 0$, and

$$\|(\alpha K)^n\| \to 0$$

by the Gelfand–Beurling formula, for all α in $\mathbb{C}$ (cf. Problem 8.6).

(b) If T has no nontrivial invariant subspace, then every nonzero vector in $\mathcal{X}$ is cyclic for T. Thus take an arbitrary nonzero x in $\mathcal{X}$ so that

$$\bigvee \{T^n x\}_{n\geq 0} = \{p(T)x \in \mathcal{X}: p \text{ is a nonzero polynomial}\}^- = \mathcal{X}.$$

Since the set $\{p(T)x \in \mathcal{X}\colon p \text{ is a nonzero polynomial}\}$ is dense in $\mathcal{X}$, it follows that every nonempty open subset of $\mathcal{X}$ meets it. Therefore, for every nonempty open subset U of $\mathcal{X}$ there exists a nonzero polynomial p such that $p(T)x$ lies in U.

(c) $U_p(x_0) = \{x \in \mathcal{X}\colon \|p(T)x - x_0\| < 1\}$ is an open set because it is the inverse image of the open ball $B_1(x_0) = \{y \in \mathcal{X}\colon \|y - x_0\| < 1\}$ under the continuous map $p(T)$ (recall: *a map is continuous if and only if the inverse image of each open set is an open set*). Thus, according to assertion (b), for every nonzero x in $\mathcal{X}$ there is a nonzero polynomial p such that $p(T)x$ lies in $B_1(x_0)$; that is, such that x lies in $U_p(x_0)$.

(d) First recall the definitions of compact sets and compact operators. A set (in a topological space; in particular, in a metric space) is *compact* if every covering of it made up of open sets has a finite subcovering; a linear transformation between normed spaces is *compact* if the closure of the image of each bounded set is a compact set. Now consider the collection $\mathcal{U}$ of all nonempty sets $U_p(x_0)$. Assertion (c) says that $\mathcal{U}$ is an open covering of each subset of $\mathcal{X}$ that does not contain the origin, and so is an open covering of $K(B_1[x_0])^-$ since $0 \notin K(B_1[x_0])^-$. But K is a compact operator so that $K(B_1[x_0])^-$ is a compact set in $\mathcal{X}$, and hence $\mathcal{U}$ has a finite subcovering of $K(B_1[x_0])^-$, say, $\{U_p(x_0)\}_{p \in \mathcal{P}}$ for some finite set $\mathcal{P}$ of polynomials. Thus, if $x \in K(B_1[x_0])$, then x lies in $U_p(x_0)$ for some $p \in \mathcal{P}$, which means that $p(T)x$ lies in $B_1(x_0)$.

(e) It is readily verified that the claimed result can be rewritten as follows. There is a sequence $\{p_k\}$ of polynomials in $\mathcal{P}$ such that, for each $n \geq 1$,

$$p_1(T) \dots p_j(T)K^j x_0 \in B_1[x_0] \quad \text{for every} \quad j = 1, \dots, n.$$

Clearly, $x_0 \in B_1[x_0]$ so that $Kx_0 \in K(B_1[x_0])$, and hence assertion (d) says that there exists a polynomial $p_1 \in \mathcal{P}$ such that $p_1(T)Kx_0$ lies in $B_1[x_0]$. Thus the above proposition holds for $n = 1$. Suppose it holds for some $n \geq 1$. Again, assertion (d) ensures the existence of a polynomial in $\mathcal{P}$, say p'_{n+1}, such that $p'_{n+1}(T)Kp_1(T) \dots p_n(T)K^n x_0$ lies in $B_1[x_0]$. But T commutes with K so that $p(T)K = Kp(T)$ and, obviously, $p(T)q(T) = q(T)p(T)$, for any polynomials p and q. Then $p_1(T) \dots p_n(T)p'_{n+1}(T)K^{n+1} x_0$ lies in $B_1[x_0]$. Therefore, the sequence $\{p'_k\}$ of polynomials in $\mathcal{P}$ obtained from $\{p_k\}$ by replacing p_{n+1} with p'_{n+1} is such that $p'_1(T) \dots p'_j(T)K^j x_0$ lies in $B_1[x_0]$ for every integer $j = 1, \dots, n+1$, and hence the above proposition holds for $n+1$, which concludes the proof by induction.

Solution 12.2. Let K be a nonzero compact operator acting on a complex Banach space $\mathcal{X}$, and let $\mathcal{A}$ be a unital subalgebra of $\mathcal{B}[\mathcal{X}]$ that has no nontrivial invariant subspace. Take any x_0 in $\mathcal{X}$ such that $0 \notin B_1[x_0]$ and $0 \notin K(B_1[x_0])^-$.

(a) Let x be an arbitrary vector in $\mathcal{X}$ and consider the set

$$\mathcal{A}_x = \{Ax \in \mathcal{X} \colon A \in \mathcal{A}\} = \bigcup_{A \in \mathcal{A}} Ax \subseteq \mathcal{X}.$$

If y_1 and y_2 lie in $\mathcal{A}_x$, then there exist operators A_1 and A_2 in $\mathcal{A}$ such that $y_1 = A_1 x$ and $y_2 = A_2 x$, and hence $y_1 + y_2 = (A_1 + A_2)x$ lies in $\mathcal{A}_x$ (because $A_1 + A_2 \in \mathcal{A}$). If y lies in $\mathcal{A}_x$, then there exists A in $\mathcal{A}$ such that $y = Ax$, and so $\alpha y = \alpha A y$ also lies in $\mathcal{A}_x$ (since $\alpha A \in \mathcal{A}$), for every scalar α. Thus $\mathcal{A}_x$ is a linear manifold of $\mathcal{X}$, and therefore $\mathcal{A}_x^-$ is a subspace of $\mathcal{X}$ (the closure of a linear manifold is a subspace). Now take an arbitrary $A \in \mathcal{A}$. If y lies in $\mathcal{A}_x$, then $y = A_0 x$ for some A_0 in $\mathcal{A}$ so that $Ay = A A_0 x$ also lies in $\mathcal{A}_x$ (once $A A_0 \in \mathcal{A}$). Then $A(\mathcal{A}_x) \subseteq \mathcal{A}_x$ and hence $A(\mathcal{A}_x^-) \subseteq \mathcal{A}_x^-$ (because A is continuous); that is, $\mathcal{A}_x^-$ is an invariant subspace for A. Since A is arbitrary in $\mathcal{A}$,

$$\mathcal{A}_x^- \text{ is an invariant subspace for } \mathcal{A}.$$

Moreover, since $I \in \mathcal{A}$, it follows that x lies in $\mathcal{A}_x$. Hence $\mathcal{A}_x^- \neq \{0\}$ for every $x \neq 0$. Therefore, if $\mathcal{A}$ has no nontrivial invariant subspace, then $\mathcal{A}_x^- = \mathcal{X}$ (i.e., $\mathcal{A}_x$ is dense in $\mathcal{X}$, which means that every nonempty open subset of $\mathcal{X}$ meets $\mathcal{A}_x$) whenever $x \neq 0$. In other words, if x is a nonzero vector in $\mathcal{X}$ and U is a nonempty open subset of $\mathcal{X}$, then $(\bigcup_{A \in \mathcal{A}} Ax) \cap U \neq \varnothing$ so that Ax lies in U for some A in $\mathcal{A}$.

(b) $U_A(x_0) = \{x \in \mathcal{X} \colon \|Ax - x_0\| < 1\}$ is open because it is the inverse image of the open ball $B_1(x_0) = \{y \in \mathcal{X} \colon \|y - x_0\| < 1\}$ under the continuous map A (compare with Solution 12.1(c)). Moreover, according to assertion (a), for every nonzero x in $\mathcal{X}$ there exists an operator A in $\mathcal{A}$ such that Ax lies in $B_1(x_0)$; that is, such that x lies in $U_A(x_0)$.

(c) Let $\mathcal{U}$ be the collection of all nonempty sets $U_A(x_0)$; that is, $U \in \mathcal{U}$ if and only if $U = U_A(x_0) \neq \varnothing$ for some $A \in \mathcal{A}$. Assertion (b) says: $\mathcal{U}$ is an open covering of each subset of $\mathcal{X}$ that does not contain the origin. Since $0 \notin K(B_1[x_0])^-$, $\mathcal{U}$ is an open covering of $K(B_1[x_0])^-$. But K is compact so that $K(B_1[x_0])^-$ is a compact set in $\mathcal{X}$. Hence $\mathcal{U}$ has a finite subcovering of $K(B_1[x_0])^-$, say, $\{U_A(x_0)\}_{A \in \mathcal{F}}$ for some finite subset $\mathcal{F}$ of $\mathcal{A}$. Therefore, if $x \in K(B_1[x_0])$, then x lies in $U_A(x_0)$ for some A in $\mathcal{F}$, which means that Ax lies in $B_1(x_0)$. (Compare with Solution 12.1(d).)

(d) Take an arbitrary operator A in $\mathcal{F}$ and consider the functions α_A, β_A and F defined in the problem statement. Since A is continuous, norm is continuous, and composition of continuous functions is again a continuous function, it follows that α_A is continuous, and so is β_A. Therefore, using the same argument and recalling that K is continuous, and also that product and finite sum of continuous functions are continuous functions, it follows that

F is continuous.

Recall that *AK* is a compact operator (because *K* is compact and *A* is bounded) so that $AK(B_1[x_0])^-$ is a compact set (by the definition of a compact operator). Hence $\bigcup_{A\in\mathcal{F}}AK(B_1[x_0])^-$ is compact (finite union of compact sets is compact). Thus $\mathrm{co}\big(\bigcup_{A\in\mathcal{F}}AK(B_1[x_0])^-\big)^-$ is compact by the Mazur Theorem, and so $\mathrm{co}\big(\bigcup_{A\in\mathcal{F}}AK(B_1[x_0])\big)^-$ is compact (since $\mathrm{co}(G)^- \subseteq \mathrm{co}(G^-)^-$ for every set *G*; and closed subsets of a compact set are compact). But $\sum_{A\in\mathcal{F}}\beta_A(Kx)=1$ so that *F*(*x*) is a convex linear combination of vectors in $\bigcup_{A\in\mathcal{F}}AK(B_1[x_0])$ for each *x* in $B_1[x_0]$. This implies that $F(B_1[x_0]) \subseteq \mathrm{co}\big(\bigcup_{A\in\mathcal{F}}AK(B_1[x_0])\big)$ (for $\mathrm{co}(G)$ coincides with the set of all convex linear combinations of vectors in *G*), and so $F(B_1[x_0])^- \subseteq \mathrm{co}\big(\bigcup_{A\in\mathcal{F}}AK(B_1[x_0])\big)^-$. Hence the closed set $F(B_1[x_0])^-$ is included in a compact set so that it is itself compact:

$$F(B_1[x_0])^- \text{ is a compact set.}$$

Outcome: $F\colon B_1[x_0] \to \mathcal{X}$ is a compact mapping.

(e) Take any vector $x \in B_1[x_0]$ so that $Kx \in K(B_1[x_0])$. If $\beta_A(Kx)\neq 0$, then $\alpha_A(Kx)>0$ (i.e., $\|AKx-x_0\|<1$), and hence $AKx \in B_1(x_0)$. Thus *F*(*x*) is a linear combination of vectors in $B_1[x_0]$ which, in fact, is a convex linear combination because $\sum_{A\in\mathcal{F}}\beta_A(Kx)=1$. Therefore *F*(*x*) lies in $\mathrm{co}(B_1[x_0])$. But $\mathrm{co}(B_1[x_0]) = B_1[x_0]$ since the set $B_1[x_0]$ is already convex. Hence

$$F(B_1[x_0]) \subseteq B_1[x_0].$$

(f) $F\colon B_1[x_0] \to \mathcal{X}$ is compact by (d) and $B_1[x_0]$-invariant by (e), and so has a fixed point by the Schauder Fixed Point Theorem. That is, there exists a vector *x* in $B_1[x_0]$ such that *F*(*x*) = *x*. Moreover, $x \neq 0$ since $0 \notin B_1[x_0]$. For each *A* in $\mathcal{F} \subseteq \mathcal{A}$ set $\beta_A = \beta_A(Kx) \in \mathbb{R}$ and consider the operator $L = \sum_{A\in\mathcal{F}}\beta_A A$, which clearly lies in $\mathcal{A}$ because $\mathcal{F}$ is finite and $\mathcal{A}$ is an algebra. But $LKx = F(x)$ so that $LKx = x \neq 0$. Thus

$$\mathcal{N}(I-LK) \neq \{0\}.$$

Solution 12.3. Let *T* be an operator acting on a complex Banach space $\mathcal{X}$. Suppose there exists a nonzero compact operator *K* in $\{T\}'$, and suppose *T* has no nontrivial hyperinvariant subspace.

(a) Put $\mathcal{A} = \{T\}'$. Since *K* is a nonzero compact operator, and since *T* has no nontrivial hyperinvariant subspace (i.e., since there is no nontrivial invariant subspace of $\mathcal{X}$ that is invariant for every operator in $\mathcal{A}$), it follows by the Lomonosov Lemma that there exists *L* in $\{T\}'$ such that

$$\mathcal{N}(I - LK) \neq \{0\}.$$

Since K and L lie in $\{T\}'$ it follows that LK also lies in $\{T\}'$ (reason: $\{T\}'$ is an algebra), and hence $LKT = TLK$. If x is any vector in $\mathcal{N}(I - LK)$, then $LKTx = TLKx = Tx$ so that $Tx \in \mathcal{N}(I - LK)$. Therefore,

$$\mathcal{N}(I - LK) \text{ is } T\text{-invariant.}$$

(b) Since K is compact and L is bounded, LK is compact (the class of all compact operators on $\mathcal{X}$ is a two-sided ideal of $\mathcal{B}[\mathcal{X}]$) and so is the restriction of it to $\mathcal{N}(I - LK)$ (the restriction of a compact operator to a linear manifold is again compact). That is, $LK|_{\mathcal{N}(I-LK)}$ is compact. But $LK|_{\mathcal{N}(I-LK)} = I$ on $\mathcal{N}(I - LK)$ so that $\dim \mathcal{N}(I - LK) < \infty$ (the identity is not compact on an infinite-dimensional space). Put $\mathcal{M} = \mathcal{N}(I - LK)$. Assertion (a) says that $\mathcal{M}$ is T-invariant, so that $T|_{\mathcal{M}}$ lies in $\mathcal{B}[\mathcal{M}]$, and also that $\mathcal{M} \neq \{0\}$. Thus, as $\mathcal{M}$ is a nonzero finite-dimensional space, $T|_{\mathcal{M}}$ has an eigenvalue (Problem 8.7). This implies that T in $\mathcal{B}[\mathcal{X}]$ has an eigenvalue (if $T|_{\mathcal{M}}x = \lambda x$ for $x \neq 0$ in $\mathcal{M}$ and some scalar λ, then $Tx = T|_{\mathcal{M}}x = \lambda x$). That is, there exists λ in $\mathbb{C}$ such that

$$\mathcal{N}(\lambda I - T) \neq \{0\}.$$

Solution 12.4. Take T, S and K in $\mathcal{B}[\mathcal{H}]$. $\mathcal{H}$ is a complex Hilbert space.

(a) Suppose $0 < \dim \mathcal{N}(S) < \infty$. Since $\mathcal{N}(S) \neq \{0\}$, take an arbitrary $x \neq 0$ in $\mathcal{N}(S)$ and consider the cyclic subspace $\mathcal{M} = \bigvee \{T^n x\}_{n \geq 0}$ so that $\mathcal{M} \neq \{0\}$ and $\mathcal{M}$ is T-invariant (thus $T|_{\mathcal{M}} \in \mathcal{B}[\mathcal{M}]$). If the nonzero subspace $\mathcal{M}$ is finite-dimensional, then $T|_{\mathcal{M}}$ has an eigenvalue (Problem 8.7). This implies that T has an eigenvalue. Thus

$$\sigma_P(T) \neq \varnothing \text{ whenever } \mathcal{M} \text{ is finite-dimensional.}$$

If $\mathcal{M}$ is infinite-dimensional, then $\mathcal{M} \backslash \mathcal{N}(S) \neq \varnothing$ (for $\dim \mathcal{N}(S) < \infty$). If $T^n x$ lies in $\mathcal{N}(S)$ for every $n \geq 1$, then $\mathcal{M} = \bigvee \{T^n x\}_{n \geq 0} \subseteq \mathcal{N}(S)$, and so $\mathcal{M} \backslash \mathcal{N}(S) = \varnothing$, which contradicts the fact that $\mathcal{M} \backslash \mathcal{N}(S) \neq \varnothing$. Thus $T^n x$ does not lie in $\mathcal{N}(S)$ for every $n \geq 1$. But $ST^0 x = Sx = 0$. Then there exists a smallest integer $n_0 \geq 1$ such that $ST^{n_0} x \neq 0$ and $ST^{n_0 - 1} x = 0$. Consider the commutator $C = TS - ST$ in $\mathcal{B}[\mathcal{H}]$.

Claim. If $\dim \mathcal{R}(TS - ST) = 1$, then $\mathcal{R}(C) \subseteq \mathcal{R}(S)$.

Proof. $CT^{n_0 - 1} x = TST^{n_0 - 1}x - ST^{n_0} x = -ST^{n_0} x \neq 0$, which lies in $\mathcal{R}(C)$. If $\mathcal{R}(C)$ is one-dimensional, then $-ST^{n_0} x = CT^{n_0 - 1} x = \alpha e$ for some nonzero α in $\mathbb{C}$, where e is the unit vector in $\mathcal{R}(C)$. If y is an arbitrary vector in $\mathcal{R}(C)$, then $y = \beta e$ for some β in $\mathbb{C}$ so that $-\beta \alpha^{-1} ST^{n_0} x = \beta e = y$, and hence $y \in \mathcal{R}(S)$. □

Now, if $\mathcal{R}(C) \subseteq \mathcal{R}(S)$, then $\mathcal{R}(S)$ is T-invariant (if $z \in \mathcal{R}(S)$, then $Tz = TSv = Cv - STv$ for some $v \in \mathcal{H}$ so that $Tz \in \mathcal{R}(S)$), and so is $\mathcal{R}(S)^-$ (Problem 1.1). But $\mathcal{N}(S^*)^\perp = \mathcal{R}(S)^-$ (Problem 2.1) so that $\mathcal{N}(S^*)^\perp$ is T-invariant. Therefore $\mathcal{N}(S^*) = \mathcal{N}(S^*)^{\perp\perp}$ is T^*-invariant (Problem 4.3), and hence $T^*|_{\mathcal{N}(S^*)} \in \mathcal{B}[\mathcal{N}(S^*)]$. Next suppose that $0 < \dim \mathcal{N}(S^*) < \infty$. Then $\mathcal{N}(S^*)$ is nonzero and finite-dimensional so that $T^*|_{\mathcal{N}(S^*)}$ has an eigenvalue (Problem 8.7), and so T^* has an eigenvalue. Thus

$$\sigma_P(T^*) \neq \varnothing \text{ whenever } \mathcal{M} \text{ is infinite-dimensional.}$$

Either $\sigma_P(T) \neq \varnothing$ or $\sigma_P(T^*) \neq \varnothing$. Therefore, $\sigma_P(T) \cup \sigma_P(T^*) \neq \varnothing$.

(b) If $\dim \mathcal{R}(KT - TK) = 1$, then it is clear that K is nonzero and T is nonscalar. Suppose T has no nontrivial hyperinvariant subspace. If K is compact, then the Lomonosov Lemma ensures that there exists L in $\mathcal{A} = \{T\}'$ for which $\mathcal{N}(I - LK) \neq \{0\}$. Now put $C = KT - TK$. Since L commutes with T, it follows that $(LK)T - T(LK) = LC$. Since K is compact and $\mathcal{N}(I - LK) \neq \{0\}$, it follows that LK is compact and nonzero (and so is $(LK)^* = K^*L^*$). If $LC = O$, then the nonzero compact LK commutes with the nonscalar T, and hence T has a nontrivial hyperinvariant subspace by the Lomonosov Theorem, which is a contradiction. On the other hand, if $LC \neq O$, then $\dim \mathcal{R}(LC) = 1$ (recall: $\dim \mathcal{R}(C) = 1$). Set $S = I - LK$ so that $TS - ST = LC$ and $\mathcal{N}(S) \neq \{0\}$. Then

$$\dim \mathcal{R}(TS - ST) = 1 \quad \text{and} \quad \dim \mathcal{N}(S) > 0.$$

Consider the Fredholm alternative. Since $1 \in \sigma_P(LK)$, it follows by Problem 8.1 that $1 \in \sigma(K^*L^*)$, and so $1 \in \sigma_P(K^*L^*)$ (since K^*L^* is compact). Therefore, $\mathcal{N}(S^*) = \mathcal{N}(I - K^*L^*) \neq \{0\}$; that is,

$$\dim \mathcal{N}(S^*) > 0.$$

Moreover, $\dim \mathcal{N}(\lambda I - LK) = \dim \mathcal{N}(\bar{\lambda} I - (LK)^*) < \infty$ for every nonzero scalar λ (since LK is compact), and hence

$$\dim \mathcal{N}(S) < \infty \quad \text{and} \quad \dim \mathcal{N}(S^*) < \infty.$$

Thus $\sigma_P(T) \cup \sigma_P(T^*) \neq \varnothing$ according to (a), and so the nonscalar T has a nontrivial hyperinvariant subspace by Problem 8.2(d), which is again a contradiction. Outcome: *If $\dim \mathcal{R}(KT - TK) = 1$ and K is compact, then T has a nontrivial hyperinvariant subspace.*

Solution 12.5. Consider the self-commutator D_T of an operator T on a complex Hilbert space of dimension greater than 1. If $D_T = O$, then T

is normal and so has a nontrivial invariant subspace (Problem 8.13). Thus suppose $D_T \neq O$. If T is quasireducible, then there is a nonscalar L in $\{T\}'$ such that rank $(D_T L - L D_T) \leq 1$ (Problem 11.1). If T is essentially normal, then the nonzero D_T is compact. Hence L has a nontrivial hyperinvariant subspace (by the extension of the Lomonosov Theorem), and therefore T in $\{L\}'$ has a nontrivial invariant subspace.

Solution 12.6. If T in $\mathcal{B}[\mathcal{H}]$ has no nontrivial invariant subspace, then every nonzero vector x in $\mathcal{H}$ is a cyclic vector for T: for every $x \neq 0$ in $\mathcal{H}$,

$$\bigvee \{T^n x\}_{n \geq 0} = \{p(T)x \in \mathcal{H}: p \text{ is a nonzero polynomial}\}^- = \mathcal{H}.$$

Thus T is a cyclic operator and, in particular, 1-multicyclic (polynomials are, of course, rational functions without poles). If T also is hyponormal, then the Berger–Shaw Theorem ensures that its self-commutator D_T is a trace-class operator, and hence compact; that is, T is essentially normal. Therefore, if a hyponormal operator has no nontrivial invariant subspace, then it is not quasireducible by Problem 12.5.

References

[1] T. Andô, *Operators with a norm condition*, Acta Sci. Math. (Szeged) **33** (1972), 169–178.

[2] R. Beals, *Topics in Operator Theory*, The University of Chicago Press, Chicago, 1971.

[3] S.K. Berberian, *Introduction to Hilbert Space*, 2nd ed., Chelsea, New York, 1976.

[4] C.A. Berger and B.I. Shaw, *Selfcommutators of multicyclic hyponormal operators are always trace class*, Bull. Amer. Math. Soc. **79** (1973), 1193–1199.

[5] C.A. Berger and B.I. Shaw, *Intertwining, analytic structures, and the trace norm estimate*, Proceedings of a Conference on Operator Theory, Halifax, 1973, Lecture Notes in Math., Vol. 345, Springer, Berlin, 1973, 1–6.

[6] S.W. Brown, *Some invariant subspaces for subnormal operators*, Integral Equations Operator Theory **1** (1978), 310–333.

[7] K. Clancey, *Seminormal Operators*, Springer, Berlin, 1979.

[8] J.B. Conway, *A Course in Functional Analysis*, 2nd ed., Springer, New York, 1990.

[9] J.B. Conway, *The Theory of Subnormal Operators*, Mathematical Surveys and Monographs, Vol. 36, Amer. Math. Soc., Providence, 1991.

[10] J.B. Conway, *A Course in Operator Theory*, Graduate Studies in Mathematics, Vol. 21, Amer. Math. Soc., Providence, 2000.

[11] J. Daughtry, *An invariant subspace theorem*, Proc. Amer. Math. Soc. **49** (1975), 267–268.

[12] B.P. Duggal, *On unitary parts of contractions*, Indian J. Pure Appl. Math. **25** (1994), 1243–1247.

[13] B.P. Duggal, *On characterising contractions with C_{10} pure part*, Integral Equations Operator Theory **27** (1997), 314–323.

[14] B.P. Duggal, I.H. Jeon and C.S. Kubrusly, *Contractions satisfying the absolute value property $|A|^2 \leq |A^2|$*, Integral Equations Operator Theory (2003), to appear.

[15] B.P. Duggal, C.S. Kubrusly and N. Levan, *Paranormal contractions and invariant subspaces*, J. Korean Math. Soc. (2003), to appear.

[16] E. Durszt, *Contractions as restricted shifts*, Acta Sci. Math. (Szeged) **48** (1985), 129–134.

[17] P. Enflo, *On the invariant subspace problem for Banach spaces*, Acta Math. **158** (1987), 213–313.

[18] P.A. Fillmore, *Notes on Operator Theory*, Van Nostrand, New York, 1970.

[19] P.R. Halmos, *Introduction to Hilbert Space and the Theory of Spectral Multiplicity*, 2nd ed., Chelsea, New York, 1957.

[20] P.R. Halmos, *Shifts on Hilbert spaces*, J. Reine Angew. Math. **208** (1961), 102–112.

[21] P.R. Halmos, *Ten problems in Hilbert space*, Bull. Amer. Math. Soc. **76** (1970), 887–933.

[22] P.R. Halmos, *A Hilbert Space Problem Book*, 2nd ed., Springer, New York, 1982.

[23] L. Kérchy, *Isometric asymptotes of power bounded operators*, Indiana University Math. J. **38** (1989), 173–188.

[24] L. Kérchy, *Unitary asymptotes of Hilbert space operators*, Functional Analysis and Operator Theory, Warsaw, 1992, Banach Center Publ., Vol. 30, Polish Acad. Sci., Warsaw, 1994, 191–201.

[25] L. Kérchy, *Generalized Toeplitz operators*, Acta Sci. Math. (Szeged) **68** (2002), 377–400.

[26] L. Kérchy, *Generalized Toeplitz operators associated with operators of regular norm-sequences*, Semigroups of Operators: Theory and Applications, Rio de Janeiro, 2001, Optimization Software, Los Angeles, 2002, 119–131.

[27] L. Kérchy and Vu Quoc Pheng, *On invariant subspaces for power-bounded operators of class C_1.*, Taiwanese J. Math. **7** (2003), 69–75.

[28] H.W. Kim, C. Pearcy and A.L. Shields, *Rank-one commutators and hyperinvariant subspaces*, Michigan Math. J. **22** (1975), 193–194.

[29] H.W. Kim, C. Pearcy and A.L. Shields, *Sufficient conditions for rank-one commutators and hyperinvariant subspaces*, Michigan Math. J. **23** (1976), 235–243.

[30] C.S. Kubrusly, *Equivalent invariant subspace problems*, J. Operator Theory **38** (1997), 323–328.

[31] C.S. Kubrusly, *An Introduction to Models and Decompositions in Operator Theory*, Birkhäuser, Boston, 1997.

[32] C.S. Kubrusly, *Elements of Operator Theory*, Birkhäuser, Boston, 2001.

[33] C.S. Kubrusly, *Asymptotically partially isometric contractions*, Semigroups of Operators: Theory and Applications, Rio de Janeiro, 2001, Optimization Software, Los Angeles, 2002, 135–144.

[34] C.S. Kubrusly, *Quasireducible operators*, Internat. J. Math. Math. Sci. **2003:31** (2003), 1993–2002.

[35] C.S. Kubrusly and B.P. Duggal, *Contractions with $C_{\cdot 0}$ direct summands*, Adv. Math. Sci. Appl. **11** (2001), 593–601.

[36] C.S. Kubrusly and N. Levan, *Proper contractions and invariant subspaces*, Internat. J. Math. Math. Sci. **28** (2001), 223–230.

[37] C.S. Kubrusly and P.C.M. Vieira, *Strong stability for cohyponormal operators*, J. Operator Theory **31** (1994), 123–127.

[38] C.S. Kubrusly, P.C.M. Vieira and D.O. Pinto, *A decomposition for a class of contractions*, Adv. Math. Sci. Appl. **6** (1996), 523–530.

[39] J. Lindenstrauss and L. Tzafriri, *On the complemented subspaces problem*, Israel J. Math. **9** (1971), 263–269.

[40] V.I. Lomonosov, *Invariant subspaces for the family of operators which commute with a completely continuous operator*, Functional Anal. Appl. **7** (1973), 213–214.

[41] M. Martin and M. Putinar, *Lectures on Hyponormal Operators*, Birkhäuser, Basel, 1989.

[42] A.J. Michaels, *Hilden's simple proof of Lomonosov's invariant subspace theorem*, Adv. Math. **25** (1977), 56–58.

[43] N.K. Nikol'skiĭ, *Treatise on the Shift Operator*, Springer, Berlin, 1986.

[44] K. Okubo, *The unitary part of paranormal operators*, Hokkaido Math. J. **6** (1977), 273–275.

[45] C.R. Putnam, *Commutation Properties of Hilbert Space Operators and Related Topics*, Springer, Berlin, 1967.

[46] C.R. Putnam, *Hyponormal contractions and strong power convergence*, Pacific J. Math. **57** (1975), 531–538.

[47] H. Radjavi and P. Rosenthal, *Invariant Subspaces*, Springer, Berlin, 1973.

[48] C.J. Read, *A solution to the invariant subspace problem*, Bull. London Math. Soc. **16** (1984), 337–401.

[49] T. Saitô, *Hyponormal operators and related topics*, Lectures on Operator Algebras, New Orleans, 1970–1971, Lecture Notes in Math., Vol. 247, Springer, Berlin, 1972, 533–664.

[50] A.L. Shields, *Weighted shifts operators and analytic function theory*, Topics in Operator Theory, Mathematical Surveys no. 13, Amer. Math. Soc., Providence, 2nd pr., 1979, 49–128b.

[51] J.G. Stampfli, *Which weighted shifts are subnormal?*, Pacific J. Math. **17** (1966), 367–379.

[52] B. Sz.-Nagy and C. Foiaş, *Harmonic Analysis of Operators on Hilbert Space*, North-Holland, Amsterdam, 1970.

[53] J.E. Thomson, *Invariant subspaces for algebras of subnormal operators*, Proc. Amer. Math. Soc. **96** (1986), 462–464.

[54] P.Y. Wu, *All (?) about quasinormal operators*, Operator Theory and Complex Analysis, Sapporo, 1991, Oper. Theory Adv. Appl., Vol. 59, Birkhäuser, Basel, 1992, 372–389.

[55] D. Xia, *Spectral Theory of Hyponormal Operators*, Birkhäuser, Basel, 1983.

[56] T. Yoshino, *On the unitary part of dominant contractions*, Proc. Japan Acad. Ser. A Math. Sci. **66** (1990), 272–273.

[57] T. Yoshino, *Introduction to Operator Theory*, Longman, Harlow, 1993.

[58] T. Yoshino, *The unitary part of $\mathcal{F}$ contractions*, Proc. Japan Acad. Ser. A Math. Sci. **75** (1999), 50–52.

Index

Additional books by Carlos S. Kubrusly

Elements of Operator Theory, ©2001, ISBN: 0-8176-4174-2

An Introduction to Models and Decompositions in Operator Theory, ©1997, ISBN: 0-8176-3992-6